Artisanal Fishers on the Kenyan Coast

Afrika-Studiecentrum Series

VOLUME 14

Artisanal Fishers on the Kenyan Coast

Household Livelihoods and Marine Resource Management

By

Jan Hoorweg
Barasa Wangila
Allan Degen

BRILL

LEIDEN • BOSTON
2009

This book is printed on acid-free paper.

Cover photo caption: 'Fishers in Watamu beaching a *mashua* vessel'
Photo by Jan Hoorweg

Library of Congress Cataloging-in-Publication Data

Artisanal fishers on the Kenyan coast : household livelihoods and marine resource management / edited by Jan Hoorweg, Barasa Wangila, Allan Degen.
p. cm. — (Afrika-studiecentrum series ; v. 14)
Includes bibliographical references and index.
ISBN 978-90-04-17808-3 (pbk. : alk. paper)
1. Fishers—Kenya—Economic conditions. 2. Fisheries—Kenya. 3. Marine resources—Kenya—Management. I. Hoorweg, Jan. II. Wangila, Barasa. III. Degen, A. Allan, 1946- IV. Title. V. Series.

HD8039.F66K422 2009
331.7'6320967623—dc22

2009022048

ISSN 1570-9310
ISBN 978 90 04 17808 3

PRINTED IN THE NETHERLANDS

Contents

Acknowledgements

This monograph is based on several years of research on resource diversification among artisanal fishers that was supported by the Netherlands-Israel Research Programme (Project NIRP-97-145-7), Moi University (Eldoret, Kenya), Ben Gurion University (Beer Sheva, Israel) and the African Studies Centre (Leiden, the Netherlands). The project consisted of eight studies in all: four major surveys and four M.Sc. studies.

Our thanks go firstly to the officials of the NIRP programme and the collaborating institutes who made the project possible: Prof. Ton Dietz, Prof. Richard Israelowitz, Prof. Mwakio Tole, Prof. Wilson Yabann, Prof. Gerti Hesseling and Prof. Leo de Haan, and the staff of the NIRP secretariat, Dr Henk Mastenbroek and Ms Loes Minkman.

We are also indebted to Mas'ad Omar Mohammed, Joseph Tunje, Nicole Versleijen and Andrew Wamukota whose thesis research was part of the project and who provided many useful insights. After completing his thesis, Joseph Tunje stayed on as field supervisor with the project.

We are grateful to colleagues who contributed their time as supervisors or advisors: Dr Dick Foeken, Dr B. Kaunda-Arara, Dr H.K. Maritim, Dr M. Muchiri, Dr Nyawira Muthiga, Dr Robert Mwadime, Dr C. Ngugi, Dr David Obura, Dr J.B. Okeyo-Owuor, Dr Marcel Rutten, Prof. Joseph Uyanga, Dr G.M. Verschoor and Dr. Enock Wakwabi.

The project was based at the Coast Environment Research Station in Malindi and we appreciated the assistance of the administrative staff there as well as the resident students: Mary Gona, Rebson Dzala, Abdirizak Nunow, Genevieve Atamba and Thomas Munyao. Mohamed Salim and Frederick Mulewa doubled as both office staff and field assistants.

We are most grateful to the fishers who acted as key informants, in particular Yusuf, Shafie and Osthaz, and to our many assistants: Rashid Anam, Mohamud Awadh, M. Boaz, Joseph Charo, Margaret Kai, Charo Karabu, Jackson Kombe Karabu, Mohammed Kheir, Gabriel Kombe, Mohammed Kombo, Sidi Mohamud, Hafith Thei Mote, Said Ali Mponye, Abdulrahman Shariff, Florence Umazi and Harrison Yeri.

We wish to thank the people who contributed to the project in less formal ways, if only through their thought-provoking discussions and enduring support: Thomas Dansforth, the captain of a sports fishing boat, Abe Gonshor, a renowned sports fisherman, John Ridgway, proud owner of a *jahazi*, Roger Benjamin of Ben-Gurion University, the late Prof. Juma Lugogo of the Coast Development Authority, Gideon Mungaro, the mayor of Malindi, Ann Robertson, the environmental conscience of Malindi, and Alice Mapenzi Kubo. Finally, we would like to thank Nel de Vink for the maps and illustrations and an anonymous reviewer of the manuscript for his helpful comments.

Abbreviations/Acronyms

ANOVA	Analysis of variance
CDA	Coast Development Authority
CT	Coastal tract
DFID	Department for International Development (UK)
EEZ	Exclusive Economic Zone
FAO	Food and Agriculture Organization
GSI	Gonadosomatic Index
H'hlds	Households
KASA	Kenya Association of Sea Anglers
K£	Kenya pound
Ksh	Kenya shilling
KWS	Kenya Wildlife Service
LS	Landing site
MENR	Ministry of Environment and Natural Resources
MR	Multiple response
MSY	Maximum sustainable yield
NIRP	Netherlands-Israel Research Programme
SL	Standard length
TL	Total length
UNDP	United Nations Development Programme
UNEP	United Nations Environment Programme
WHAT	World Humanity Action Trust
WIOMSA	Western Indian Ocean Marine Science Association

Glossary

Baruti	Explosives
Boda boda	Bicycle taxi
Bunduki	Spear gun
Chapati	Flat bread made from wheat flour
Dau	Medium-sized vessel
Dhow	General term used for medium-sized and large vessels
Fundi	Craftsperson
Hori	Canoe made from planks
Jahazi	Large-sized vessel
Jarife	Floating net
Juya	Beach seine
Kaskazi	Season of north-east monsoon
Kimia	Sardine net
Kitoweo	Part of the catch kept for domestic consumption
Kusi	Season of south-west monsoon
Makuti	Roofing material from the leaves of coconut trees.
Malema	Fish trap
Mashua	Large-sized vessel
Matatu	Small van, operated as means of public transport
Mavumba	Pounded fish emitting strong smell
Mchupa	Fish poison
Mshipe	Hand line
Mkonjo	Metal rod
Mpweke	Gill net
Mtumbwi	Dug-out canoe
Mzimu	Sacred place near the sea
Mzungu	White man
Ngalawa	Outrigger canoe
Sadaka	Traditional marine ceremony
Shamba	Farmland
Tajiri	Fish trader and entrepreneur
Talbisi	Side matting for vessels
Tanga	Sail
Ugali	Stiff porridge made from maize flour
Uzio	Fishing fence

List of tables

List of figures

List of boxes

List of maps

Map 1 Kenyan coast with Marine Protected Areas

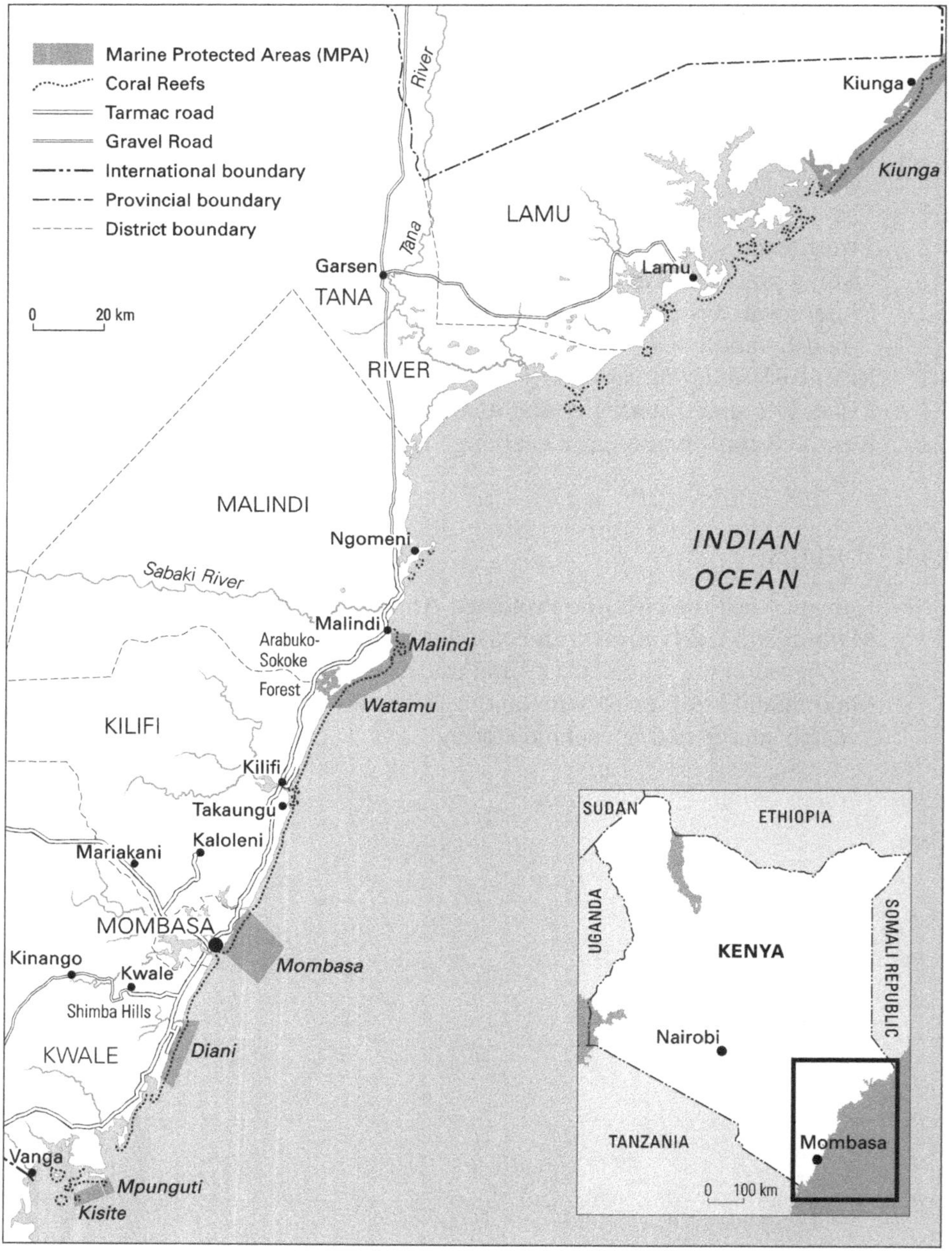

1

Sea fisheries in Kenya

Background history

The Kenyan Government, in 1928, invited Dr. Cecil von Bonde, Director of Fisheries in South Africa, to conduct a survey on the potential of sea fisheries in Kenyan waters, and to identify commercially trawlable areas. The mission did not start well. Von Bonde complained that "unfortunately upon my arrival I found that my requirements had not been anticipated, with the results that much valuable time was wasted" (Kenya 1928: 1). But that was not all, since it turned out that for eventual trials with a beam trawl only "a boat of about 12 ft. had been placed at my disposal" (Kenya 1928: 6). This was a most unsatisfactory state of affairs. But, not to be defeated he next managed to obtain the Customs motor launch, however, this vessel required the "superhuman effort of half a dozen men to get a small bean trawl aboard after it had been shot" (Kenya 1928: 6). The Fisheries Survey Committee gave permission to hire the steam vessel *Mvita* but, again, there were difficulties getting the trawl aboard and the boat could only be used "in a dead calm sea and even then only with moderate success" (Kenya 1928: 6). Von Bonde reluctantly abandoned this part of the survey but from examination of sea charts he still identified four areas that were potentially suitable, notably waters off Malindi, Kilifi, Gazi and Vanga. He was quite empathic in recommending "an intensive survey of these regions by means of a suitably equipped vessel" (Kenya 1928: 8). With this parting statement he left the undoubtedly exasperated colonial officers behind. The latter, however, somehow had the last word because in official documents 20 years later, it was stated that Dr. von Bonde had concluded that "There was no possibility for developing a trawling industry" (Kenya 1950a: 28). It is not clear who was responsible for this distortion but it appears as if the staff of the Game Department was less than keen to welcome industrial fishing on its patch. Anyway, for the moment, the notion of trawling fisheries was laid to rest because of the financial constraints at the time due to the depression in the thirties (Kenya 1950a).

Two years earlier, in 1926, the Game Department had appointed a Fish Warden who was given the responsibility of angling and trout conservation. Trout was not native to the country but had been introduced in 1905 in the River Gura on the eastern slopes of the Aberdares by Major Ewart Grogan[1] (Kenya 1938). The focus on trout fishing, a sporting activity reserved for white settlers, remained for quite some time. Shortly before the outbreak of World War II, a report appeared on the control and development of fishing in Kenya, which was still solely devoted to trout fishing. The report covered, among others, details of more than 80 individual trout rivers and their ecology, the angling clubs, the trout wardens and native scouts, and the revenues from licenses and expenditures. No mention was made of the marine or freshwater fisheries of the African population (Kenya 1938).

It was only after World War II that attention was given to 'local' fisheries. Then, the Fish Section of the Game Department was expanded so, that in addition to trout conservation, it assumed responsibility for coast fisheries and the exploitation of the resources of inland waters (Kenya 1947). At the time, the narrow coastal strip was still under the suzerainty of the Sultan of Zanzibar, and the Colonial Office decided to promote sea fisheries in an interterritorial context. A marine research scientist was appointed for the East African Coast in 1949 and he was stationed in Zanzibar – to review the possibilities of commercial fishing (Ommanney 1955). Public interest in the fifties, if any, was more with commercial sea adventurers judging from the popular accounts of the slaughter of sharks and sea turtles in Seychelles and Somali waters (Travis 1959, 1961, 1967).

In the meantime, in Kenya, an Assistant Fish Warden for coastal fisheries had been appointed in 1947 and he was stationed in Malindi. The first officer resigned within the year, but in 1948 he was succeeded by T. Allfree who remained in this post until 1962. His first tasks were the improvement of the coast fisheries together with development of marketing and distribution, improvement of native fishing methods, and collection of fishery statistics (Kenya 1950b: 28). During 1949, with great effort, he managed to visit "every village or centre of importance on the coast, at least once, ... to record the number of men and craft employed, and the catches, together with the methods used" (Kenya 1950b: 33). Having familiarized himself with the inshore fishery, the warden came out on the side of the local fishers:

> I would like here to dispel the myth that Kenya Coast fishermen work less when fishing and prices are good ... It is where the return from his day's work reaches a low level due to bad fishing conditions or poor price that he loses heart, and only catches enough for his immediate requirements, or does not go to sea. ... The trouble with the Kenya sea fishing industry is not the laziness of its fishermen, nor the so-called primitive methods employed but the insufficiency of numbers of fishermen to supply an ever increasing demand for fish (Kenya 1952: 47).

He arrived at an estimate of over 2,500 fishers with 1,019 vessels for all of the coast and an annual catch of 2,249 tons (Kenya 1950b: 33, 35). But he also mentioned a number of typical problems that hindered the development of the sea fisheries:

> The industry has difficulty in obtaining adequate supplies of gear and tackle and was having to pay excessively high prices (Kenya 1950b: 31).
>
> ... all craft are on a share basis, ... the stealing of a portion or the whole of the catch by the skipper and crew has, ... stopped the flow of capital into sea fisheries (Kenya 1950b: 34).
>
> Fish poisoning continues along the Kenya coast in those areas occupied by the Wanyika tribes. The effects of these poisonings has to be seen to be believed; in one case at a mile distant the shores of a creek were seen to be white with dead fish (Kenya 1952: 48).

During the fifties, sea fisheries started to receive more attention from the Government while trout fishing suffered a severe decline because of the movement restrictions at the time of the Mau-Mau Emergency. In 1958, when things were more or less back to normal, the annual fisheries report covered the fisheries of Nyanza Province (rivers and dams), the fish culture farm (mostly concerned with stocking of fish ponds, in particular tilapia) and trout fishing. Sea fisheries received by far the most attention and comprised more than half the report (Kenya 1959). This remained so, even when the yields from Lake Victoria suddenly surpassed the coastal captures in 1962 (Kenya 1963). The main bottlenecks of the sea fisheries, however, that were mentioned time and again were the shortages of the marketing and distribution system:

> At present the industry suffers from short price depressing gluts, and a lack of cold storage (Kenya 1950b: 29).
>
> ... the fish marketing trade will only buy on alternate days during the height of production, and then give such a price that it does not pay to catch (Kenya 1951: 28).
>
> Incredible as it might sound, ... [one] ... problem that the fish distribution trade has is to dispose of its large fish ... with the exception of Kingfish, not one single pound of fish is sold by the pound as fillets or cutlets (Kenya 1954: 26-27).
>
> ... it is essential to the whole industry that a plant be provided capable of processing and maintaining fish in cold store ... Ordinary cold stores are not suitable for this purpose ... No cold store in the country is capable of this (Kenya 1954: 26).

Finally, in 1959, cold storage facilities were constructed in Mombasa and a large space was leased by a fish marketing firm in Nairobi (Kenya 1959). Regretfully, this did not mean that this particular constraint had been overcome because, already the next year, the Kilindini Port Cold Store was closed to locally caught fish due to the greatly increased volume of perishable foods in transit through the port (Kenya 1960). In 1962, it was once again repeated that "it was unlikely that substantial progress would be made in the sea fisheries without Government intervention in marketing" (Kenya 1962: 10).

In 1960, a start was made with a Loans to Fishermen Scheme that was intended chiefly for the purchase of new equipment. During the first round, 79 loans were approved which amounted to £ 4,944 in total (Kenya 1962; Martin 1973). Since the loans were made from a revolving fund, it could only function properly if repayments were made. This turned out to be the major weakness in the scheme as there were high defaulting rates in the years to come (Kenya 1966). Still, in general, the loan scheme was considered to be successful because it enabled fishers to take advantage of the modern equipment that was being tested by the department (Kenya 1968). By 1970, the scheme had lost most of its funds and was discontinued, but, after reorganization, loan applications were invited again in 1977 (Okidi 1979).

From 1948 to 1958 there was a large expansion in the quality and the value of fish sold mainly because of the successful introduction of modern equipment and new methods of catching fish (Martin 1973). The hand lines of local cotton thread were replaced by nylon lines, which were stronger and more efficient. Particularly successful was the introduction of a blue-grey nylon shark net, which was almost invisible to the sharks. Shark catches in Malindi multiplied fourfold. During the initial years, there was little concern for any environmental consequences of the improved fisheries, which was understandable given the small number of fishers and their modest production:

> The duty of the Department is to foster the development of the fishing industry in all its aspects. With few exceptions, the Department is not concerned with the conservation of fish, for there is no evidence that man's efforts from our coast have reached a stage at which they would endanger the Colony's marine assets (Kenya 1955: 20).

Still, it did not take long before adverse effects came to be noticed in respect to certain species and certain areas. There were signs that all was not well:

> The Kenya shark fishery ... is declining rapidly, and is reaching the point at which it is no longer profitable (Kenya 1960: 11).
>
> Turtles were incorporated into the Wild Animals Protection Act in Kenya to protect them ... [from extinction] ... However, little development can be envisaged for the fishery and the Department's efforts are directed mainly at enforcing this legislation (Kenya 1964a: 11).
>
> Of great concern has been an alarming increase in the collection of live coral, shells and reef-fish for exports overseas (Kenya 1968: 21).

These and other concerns led to the start of the first Marine Protected Areas on the Kenyan coast in 1962:

> It has been apparent for some years that there is a need to reserve limited areas of the Kenya coast so as to ensure the preservation of some of the very colourful coral fish which are a delight to the eye and a major attraction to overseas visitors. The first of these reserves, at Watamu, ... has been an unqualified success. ... Within two years from the creation of the Watamu reserve, the effect on fish life has been

> most marked. A second reserve has already been delineated about four miles south of Malindi, and legislation ... will be published in 1964 (Kenya 1964a: 17).

The early sixties saw important organizational changes. In 1962, the Fisheries Department became independent of the Game Department to which it had formerly belonged. The Coast Fisheries Department itself was reorganized in 1964 with the transfer of the Provincial Fisheries Development Officer and the majority of the staff to Mombasa; Malindi remained a substation with a skeleton staff. The First National Development Plan, that year, contained 3 pages on fisheries with separate mention of inland fisheries, coastal inshore fisheries and plans for deep-sea fisheries. The need for marketing improvements was recognized again. In regard to the financial reservations, an amount of £10,000, which was destined for the inshore loans fund, was allotted for the plan period (Kenya 1964b).

The next Development Plan, 1970-1974, was more ambitious and reserved K£ 224.000 for inshore fisheries during the five-year period. Government support consisted of setting up the Kenya Inshore Fisheries company to exploit the crustacean resources, improve marketing facilities and assist the fishers loan programme (Kenya 1969). Plans for deep-sea fishing were shelved again, waiting further studies. The Development Plan of 1974 identified three principal development programmes, namely fisheries research, training and extension, and development of fisheries. The budgetary reservations were K£ 1.6 million for recurrent expenditure and K£ 1.0 million for development expenditure (Kenya 1974).

T. Allfree, the coastal fish warden, also studied the activities of the fishers from a sociological angle. Although it was a modest research programme, results were quite informative at the time. Data were collected between 1949 and 1955 on individual catches, the time spent on fishing and non-fishing activities as well as incomes of 11 Malindi fishers (Kenya 1950b, 1956). In the mid sixties, FAO (1966a) commissioned a study on the development prospects of fisheries in East Africa. This resulted in a most thorough report that covered sea fisheries and inland fisheries of 15 countries and presented figures on the marine catches in these countries. East Africa, it was concluded, was not particularly well endowed with fishing resources. The main features of the marine fisheries at the time were that little fishing occurred outside the continental shelf and that a wide variety of species were caught but few in large quantities. Underlying factors were that most fishing crafts were dugout canoes and simple plank vessels, that many fishers were employed only part-time or only occasionally in fisheries, and that during the south-west monsoon, the weather conditions were adverse throughout the region.

Subsequently, several more marine resource surveys were done in Kenyan waters (Venema 1984). In fact, the Fisheries Department mentioned 1965 as "a year of intensive activity ... providing counterpart services to the overseas ... experts who have been carrying out the two surveys of the Coast fishery" (Kenya 1967: 15). This must have been in reference to the studies on deep-sea and long-line fishing that were later followed by surveys on existing tuna and crustacean resources (FAO 1966a, 1966b, 1969, 1971). Further surveys covered the Western Indian Ocean and the trawlable coastal waters of Kenya (FAO 1979; FAO 1983). Results indicated a moderate fishing potential for Kenyan waters, giving the optimum inshore fish production of about 20,000 ton per year. Reasonably equipped shrimp trawlers should be able to land three to four tons of marketable crustaceans per day (Odero 1984).

In the meantime, Charles Okidi, a researcher at the Institute for Development Studies in Nairobi, realized the need to broaden the interest from technical aspects of fisheries to the management of coastal and offshore resources in general. In 1977, he organized a workshop that covered legal matters, coastal tourism and marine pollution, among others (Okidi & Westley 1978). Independently, Okidi (1979) drafted a policy paper on marine fisheries, reviewing the economic value of this sector in the coastal economy, the constraints experienced by the fisheries department, the future of fisher cooperatives and the problems with the loan scheme. Most of the report concerned artisanal fisheries but attention was also given to commercial fishing and even to the intrusion of foreign long-distance fleets, despite the existing scarcity of information.

Artisanal fisheries

The term fishery generally refers to the industry or occupation of catching, processing and selling finfish, shellfish and other aquatic organisms. Fisheries comprise marine fisheries, inland fisheries and aquaculture and they can be described based on the volume, purpose and intensity of the fishery. Thus a fishery can be referred to as industrial or artisanal and, depending on level, size and purpose, as small-scale or large-scale commercial. A fishery can also bear the name of the catch such as tuna fishery, cod fishery, perch fishery, shrimp fishery or lobster fishery. Location can also be used as in the reference to lake fishery, riverine fishery, marine fishery, estuarine fishery or inshore fishery. Still, others may refer to a fishery by the dominant gear or method used for catching the fish so that there is, for instance, a gill net fishery or trawl fishery. A combination of names is also common.

The term artisanal fisheries typically refers to traditional fisheries that involve households (as opposed to companies) using relatively small amounts of capital and small fishing craft, making short trips and staying close to shore with the catch destined mainly for local consumption (Charles 2001; FAO 2004a). Artisanal fisheries are often referred to as small-scale fisheries in the existing literature but the term artisanal is preferred here because of the link with household livelihoods.[2] All fishers, whether using traditional fish traps, fish spear gun, hand line and hooks, nets or diving and collecting, belong to artisanal fishery.

The importance of fisheries in the economies of many countries is well documented since FAO started its annual reviews of the state of the world fisheries in 1957. Global production from captured fish and aquaculture has been estimated to provide more than 15% of total animal protein (FAO 2002) and even more in poor countries and coastal regions (Béné, Macfadyen & Allison 2007). Since fish is a relatively cheap source of food in most countries, it is particularly important for the poorest sections of the population (Walmsley, Purvis & Ninnes 2006). The total global yield of fish in 2004 was estimated at 140.5 million tons of which about a third was from aquaculture. Of the 95.0 million tons of fish caught, 85.8 million tons were from marine waters (FAO 2007).

Artisanal fishers exploit most of the coastal fisheries of Sub-Sahara Africa. Distant fishing fleets from Europe and Eastern Asia harvest most of the lucrative oceanic fish. The Western Indian Ocean and the South-east Atlantic, covering most of Sub-Saharan Africa, had an estimated production of about 7.5 million tons of fish in 2004 (FAO 2007: 30). The total number of full-time, part-time and seasonal fishers in Africa, excluding fish farmers, was estimated at 2.73 million in 2005 (FAO 2007: 23). More than 90% of the world fishers belong to the small-scale sector and, including the people involved in the processing and other related industries, this may increase the number of people dependent on fisheries for their livelihood in Africa to close to 10 million people (Béné *et al.* 2007).

Sea fisheries are characterised by uncertainty. Fishers are poorly equipped for the dangers of the sea, are dependent on middlemen and ship-owners and have to deal with fluctuating incomes (Charles 2001). Their gear have to be adapted and different fishing techniques have to be mastered to match the seasonal changes and the large number of species. Furthermore, fishers are usually politically underrepresented because of their regular absences due to long periods at sea and various other reasons (Kenya 1956).

Artisanal fishers enter agreements to form crews as well as larger interest groups to increase catch, reduce risk and defend their general interests. Fishers belonging to a crew are rarely paid a wage, but usually receive a portion of the catch. The catch has to be divided among the crew and shared with the owners of the vessel and gear. Generally, shares of the catch are distributed according to labour and capital that has been contributed (Acheson 1981). Crew relations are commonly egalitarian in that captains usually have only limited authority. The composition of the crew is diverse and may consist of kinsmen, friends or non-kinsmen. Sometimes younger children start fishing with their fathers, uncles or brothers and they may join other crews later.

In some fishing societies, the boat and gear are owned by members of the crew. However, fishing gear and vessels can also be owned jointly by lineage members, co-operatives, fisher organizations, village committees or be the individual property of businessmen. In Kenya, ownership of gear sometimes rests with the *tajiri*, a person involved in marketing the fish at most landing sites. They are "often older, former fishers, or part owners of the boats fishers use" and can also provide credit to fishers (Glaesel 1997a: 58; Martin 1973).

Fishers also organize themselves to try and reduce risk. Klein (1999) mentioned the strong sense of communal responsibility among fishers on the Nigerian Coast and that fishing retains features of a collective enterprise in many villages. Hinrichsen (1998) described a fisheries co-operative in Vanga

(Kenya), which helped stabilize incomes of fishers. Lopes *et al.* (1997) regarded collective action as a way to deal with risk. The bond of companionship and brotherhood provides security to fishers. In contrast, Knowles (1987) reported that fishers of Pate (Kenya) had minimal contact with the community, although fishing played an essential role in the local economy. Another way for fishing societies to cope with uncertainties and risk is through traditional rituals and magic, often showing a concern with cleanliness and hygiene (Tunje & Hoorweg 2003).

Fishers can also deal individually with risk, uncertainty and competition. Through skill, capital management, innovation and technical change, a fisher can limit risk and deal with competition. Skills are an important individual asset and often there is a reluctance to share information on fishing practices and movements of fish stock among fishers. Access to new and more effective fishing gear and vessels are also important for fishing success. However, fishers often have to reject innovations or do not have access to them due to lack of money and/or geographical distance. Innovations are likely to be rejected when they are not profitable or incompatible with existing cultural patterns. In addition, innovations may increase yield only when fish are abundant (Acheson 1981). Fishing success in the long-term is not only linked to skills but also to the ability to handle and invest money generated during the times of good catches.

Fisher incomes and poverty

Fishers in developing countries are often regarded as poor both in absolute and relative terms. Béné (2003, 2004) distinguishes two types of explanations for poverty among fishers. The first explanation emphasizes the low levels of natural resources and fishing industry related factors. Fishers are poor mainly as a result of the open access of the resource and the influx of people in the fishing sector, which leads to overexploitation. The second explanation relates the fishery sector to other sectors of the economy and argues that there should be a wage equilibrium between the fishery and non-fishery sectors. But often this is not the case as fishers usually live in remote areas and lack of alternative employment is one of the key factors contributing to low standards of living in fishing communities.

Despite the fact that most fishing activities in East African waters are artisanal and therefore of small scale, nearshore fisheries are being over-exploited along most of the mainland coast (Hinrichsen 1998). In Kenya, the coastal environment and its valuable resources are increasingly under pressure from human settlement and related developments. Faced with dwindling resources and more and more competition, not only from fellow fishers but also from tourism and human settlement, fishers have little choice but to adjust to the changing circumstances. One alternative is to fish more intensively, for example, by investing in vessels and gear, but this is beyond the means of most fishers. Another way to cope with uncertainty lies in livelihood diversification, that is, engaging in economic activities other than fishing (Allison & Ellis 2001; Tvedten & Hersoug 1992; Béné *et al.* 2007). Already, most fishers do not set out during the windy and rainy season when waters are rough, and they often use this period for other activities.

The livelihood concept gained rapid acceptance in the early nineties among researchers after the term was made popular by Chambers (1987) in a series of publications on sustainable livelihoods. Livelihood refers to a source of maintenance and a means of living and it may either refer to an individual or a household. Researchers who were critical of the development policies of international organizations such as World Bank and International Monetary Fund found the livelihood concept a useful tool to study poverty among households in third-world countries. Livelihood became the unifying concept for studies on poverty, inequality, risk and insecurity from different disciplines. Often the term was used in combination with 'strategy' to emphasize that people are active agents with their own perspectives, strategies and judgments in making a living and dealing with the environment (Kaag *et al.* 2004: 68).

Livelihood diversification is a widespread survival strategy of rural households in Sub-Saharan Africa. Many studies on household diversification focused on farm households and pastoralists but there has been little attention on fisher households. Diversification is expected to improve household livelihoods, first of all, to increase income and/or to better the income spread. Some researchers, however, argue that specialization offers better opportunities to improve incomes under certain conditions (Scoones 1998; Ellis 1999). It is important, though, to distinguish between diversification at household level and diversification at individual level, that is, 'earner' diversification where the household has more than one income earner, and 'activity' diversification where the head of the household has income from more than one activity (Ellis 1998; Woodhouse 2002). It should also be noted that trends can work in reverse, that is, people involved in, for example, agriculture or wage employment may decide to start fishing in order to diversify their incomes.

The role of women in fisheries has long been underestimated but has recently started to receive more attention (Williams *et al.* 2002; Horemans & Jallow 1997b). Although there are few societies where women actively fish, women are commonly engaged in fish processing and fish marketing and, thus, link production and consumption. Women's activities are often small scale and their incomes are less than their male counterparts. They suffer constraints such as lack of credit and training, inadequate markets and transport problems that deny them access to distant markets. Nonetheless, women have an important role in household diversification because they are often engaged in other economic activities such as food selling, handicrafts production and day labour (Touray 1996).

Resource conservation

Poverty has long been associated with overexploitation of natural resources, and it has been widely assumed that income improvements are needed among local populations to lessen the pressure on these resources. However, this expectation has often met with disappointment (Ellis 2000). Intensive fishing reduces the abundance of target species and, as a result, fish stocks can be depleted and, in the case of overexploitation, be threatened with extinction. Fisheries may also selectively remove larger fast-growing individuals and thus change the characteristics of fish populations (Pauley *et al.* 2002). Marine waters share aspects of common property and open access. Since everybody can freely use the existing natural resources, the individual tries to maximise his profits while the community has to share the costs. According to Bulte (1997: 55), "fishermen have no incentive to take into account the value of fish left in the sea or the shadow price of the resource." Abdullah *et al.* (2000) are indeed of the opinion that many of East Africa's valuable resource areas could, until recently, be defined as common property with open access. The assumption here is that access to the sea and harvesting of marine resources lack regulation but the reality is often more complex. Lopes *et al.* (1997), for example, denied the claim that there is open access to marine resources in their study from Mozambique.

Fishers have ways to control who is allowed to fish and how (Ostrom 1999), with two important types of control existing across fishing communities. Firstly, there are informal rules on the gear that can be used and how it can be used. In fact, there are examples of fishers being chased away by the fishing community for using destructive gear (Tunje 2000). Secondly, regulations exist in many countries to limit the access to fisheries, although the political strength needed to enforce these regulations is often missing (Alidina 2005).

Acheson (1981) described how boats that reach the fishing grounds first have temporary usufruct rights, and in this way fishers take turns in exploitation and exclude outsiders. McClanahan *et al.* (1997) described indigenous ways of conservation and regulations concerning access to fishing grounds at the Kenyan Coast. Glaesel (1997a) described sacred areas on land as well as in the sea that are identified by elders through visions in dreams and where fishing is not allowed. Sacred areas often include ecologically important habitats, for example fish breeding grounds. Spirit-based beliefs have shaped community practices to include 'modern' methods of fisheries management such as closed seasons, limited access, size restrictions and protected areas. The sacred areas and access regulations by elders in Mozambique have been eroded because community ownership of resources was not recognized in colonial and post-independence legislation (Chilundo & Cau 2000).

Most authors agree that the role of traditional access regulations and indigenous ways of conservation has lessened. Traditional fishing rights that used to exist in many parts of the world have largely lost their effectiveness because of lack of legal recognition, introduction of modern technologies, lack of community cohesion and lack of power to control new entrants (WHAT 2000). However, in most countries there are statutes, which impinge directly or indirectly on the coastal and marine environment. Initiatives of local community members, government sympathisers and external agents have led to various legal measures and public regulation alternatives (Horemans & Jallow 1997a). But, "national, coastal and environmental legislation are often in disagreement and have resulted in overlapping and sometimes conflicting mandates in dealing with coastal and marine issues" (Obura 2001: 1276). The legal framework in respect of coastal management in Kenya has been reviewed by Ochieng *et al.* (2001).

The most commonly used device is to limit the number of fishers through licenses. However, licenses for fishers, boats or gear in a specific area does not necessarily create an incentive for the fishers to limit fishing effort. This objective is better realised with catch quotas. Output controls limit take-off and so limit the catch of a fish species. However, introducing a quota system may result in dumping by-catch when a fisher does not have a quota to cover the latter. Another option used is one of gear restrictions, that is, limiting the use of particular fishing equipment by either type or amount. In this way, a drawback of licensing is also covered, and technological change is accommodated. Yet, another option to limit the pressure on fishery resources is that of closed seasons. Two types of closed seasons exist; (i) periods of the year are closed for certain species, and (ii) the season is closed when the catch rate declines to a predetermined point. Closed areas, such as Marine Protected Areas, are another option and have similar effects to closed seasons of the first type (WHAT 2000; Charles 2001). The five methods mentioned so far: licensing, quotas, gear restrictions, closed seasons and closed areas all require supervision. Here problems emerge because of the generally poor enforcement of regulations.

A final possibility lies in attempts to establish ownership: forming co-operatives to strengthen social pressure and to rule out 'rape, ruin and run' behaviour. Hauck & Sowman (2001) reviewed the role of co-management as a solution to overexploitation, illegal use of gear and conflicts between conservationists and local communities in South Africa. Makoloweka & Shurcliff (1997) described a community-based approach in Tanga to address declining catches, use of destructive fishing techniques, mangrove cutting and coastal erosion. These authors emphasised that, besides the local fishing community, the regional and district government officers and extension workers should also be incorporated. Many authors agreed that more discretion should be left to individuals, local organizations and agencies to adapt their conduct to a spirit of public responsibility (Dubbink & Vliet 1996; Plummer & Fennell 2007).

Many conflicts in fisheries are the result of a sectoral approach to the management of coastal and marine resources, resulting in poor government policies. The coastal zone is used extensively by many groups with an increasing number of activities which are often not compatible and which often result in conflicts (Masalu 2000; Charles 2001). Others have argued the need of an integrated policy framework concerning the use of coastal resources (Okemwa, Ruwa & Mwandotto 1997). Without one regulatory body to address coastal management issues, it is hard to find adequate solutions for conflicts that occur.

Fishing on the Indian Ocean coast

Kenya belongs to the bottom 20% of countries in the world in terms of economic and human development. About half the population is termed poor by national standards and the poorest regions of the country are North-Eastern, Nyanza and Coast Province (Kenya 2001). The reasons for poverty in Coast Province include unfavourable climate, poor agro-ecological conditions, lack of employment opportunities and low level of education (Hoorweg, Foeken & Obudho 2000). Marine fisheries are one of the few economic activities found all along the coast and the number of fishers is steadily increasing.

The coastal and marine environments of Kenya, however, are threatened by naturally occurring processes, growing subsistence needs of the coastal population, and increased economic activities in general (Hoorweg 1998). Examples of natural processes are coral bleaching, sea level change and beach erosion. Growing subsistence needs are behind the over-harvesting of mangrove trees, illegal shell collection and intensive fishing. Increased economic activities result in increases in sewage and waste disposal from tourist hotels, industrial pollution of waters near Mombasa, and siltation at river estuaries as a result of soil erosion upcountry. The first national environment plan already listed many of these issues but efforts at 'integrated coastal management' since then have been limited to the Mombasa and Diani areas (MENR 1994; Okemwa *et al.*

Box 1.1 The legal framework of the marine fisheries

The *Fisheries Act* (Kenya 1991) was first enacted in 1989 and defines the administration of fisheries in the country, including fisheries management and fisheries development. The bill covers the registration of fishing vessels, the licensing of fishers, fishing offences and their enforcement and special provisions regarding marine mammals and loan schemes to fishers. A license is required for any type of fishing and each vessel requires a certificate of registration that has to be renewed annually. The act is important for artisanal fishers as it prescribes licenses, sets conditions for certain fish species, and defines the seasons and breeding areas that are off-limits. In 1991, The *Fisheries (General) Regulations* were attached, concerned mainly with the administrative, licensing and enforcement provisions mentioned earlier. Other provisions concerned the protection and conservation of fishery waters. The act contains a long list of licenses required for different purposes such as aquarium fishes, oysters, sport fishing, trout, fish processing, fish trading, crustaceans and bêche-de-mer. The *Foreign Fishing Craft Regulations* (also 1991) comprises the licensing of foreign fishing vessels, the control of vessels in Kenyan waters, marine research and other miscellaneous provisions.

The *Wildlife (Conservation and Management) Act* (Kenya 1989) was first posted in 1976, with subsidiary legislation in 1985 and 1989. The act was evidently written with land parks and land animals in mind although the sea areas can also be regarded as land areas for purposes of the Act. Extractive activities, without exception, are forbidden in Marine Parks, however, certain activities, such as fishing using traditional methods, are allowed in Marine Reserves. There is a revised *Wildlife Bill* (2007) being prepared with a section which stipulates the establishment of marine protected and community marine conservation areas. These areas have to be managed by an approved management plan prepared after consulting the communities concerned and relevant lead agencies. Marine zones may be identified for different purposes: (non-)extraction of marine resources; protection of nesting, breeding and foraging areas; no-take areas in respect of fisheries; conservation of marine resources; and certain specified human activities.

Two other acts can also be mentioned: The *Maritime Zones Act* (Kenya 1999a) defines the territorial waters (12 nautical miles seawards and all of Ungwana Bay) and the Exclusive Economic Zone (200 nm seawards) adjacent to the territorial waters. Kenya exercises sovereign rights with respect to the exploration, exploitation and conservation of the natural resources in the zone. The provisions of the *Fisheries Act* also apply to the EEZ. The *Environmental Management and Coordination Act* (Kenya 1999b) has a section (55) that allows for the declaration of a protected coastal zone but only after an environmental survey has been made and after the preparation of a coastal zone management plan. Further clauses concern the exact content of the survey and management plan; the offence of pollution and environmental damage including pollution from land based sources, vessels and aircraft, mining equipment and artificial islands.

1997; CDA 1996, 2001). Furthermore, various sector-based legislations are concerned with the marine fisheries, namely the Fisheries Act, the Wildlife Act, the Maritime Zones Act and the Environmental Management and Coordination Act (see Box 1.1).

Kenya has about 600 km of marine coastline and a number of Marine Protected Areas, which have been instituted successively since the early sixties (Map 1; p. xii). The first Marine Park (called Coral Garden Fish Reserve) was established in Watamu in 1962 followed by a Park in Malindi in 1964 (Kenya 1964a, 1966). In 1997, along the entire length of the Kenyan Coast, the protected areas covered more than 100 km of seafront. They comprised four Marine National Parks with a total area of 54 km^2 (Malindi, Watamu, Kisite and Mombasa) and six Marine National Reserves with a total area of 781 km^2 (Malindi, Watamu, Mpunguti, Mombasa, Kiunga and Diani; WIOMSA 2007).

For fishing purposes, Kenyan waters can be divided into three zones. The first extends five nautical miles seawards and fishing in this zone is for artisanal and sport fishers only. Prawn trawlers, however, are often accused of fishing illegally in this zone. Artisanal fishers may venture further out but most of their activities occur within the five nautical miles. Sport fishers, however, often set out further seawards.

The professional sport fishers in Kenya have recently formed the Kenya Association of Sea Anglers (KASA) with about 35 charter boats. In addition, there are, perhaps, another 35 charter and private boats. KASA members are required to submit records of their catches, which is not the case with the non-member boats, although the catches of the latter are likely to be much lower than those of the professional charter boats. In 2002/2003, the total catchweight reported by sport fishers was 235,308 kg (Wright 2008) with the largest landings in Malindi (46%) and Watamu (37%). The main species caught were tuna (48%), tiger shark (10%) and sailfish (10%), and smaller quantities of billfish (black marlin, blue marlin, striped marlin, broadbill), shark (hammerhead, mako, tiger, other) and gamefish (barracuda, cobia, dolphin, kingfish, trevalley, wahoo) (Wright 2008).

The second zone is between five and twelve nautical miles seawards and together with the first zone constitutes the territorial waters. This is the zone where the prawn trawlers are allowed to operate against payment of an annual license fee (Ksh 22,800). Currently there are seven vessels active which are all Kenyan registered. The trawling season is open from March 1 to October 31 and the average annual catch totals 237 metric tons (Kochey 2008).

The marine landings in 2005 of artisanal fishers, sport fishers and prawn trawlers combined were estimated at 7,616 metric tons with a value to the fishers of Ksh 587.2 million or US$ 8.1 million (Kenya 2007a).

The third zone exists between 12 and 200 nautical miles offshore and is the Exclusive Economic Zone (EEZ). Commercial fisheries are permitted here but fishers are actually instructed to respect a 15 nautical mile zone. The potential yield of the EEZ has been estimated to be as high as 150,000 tons (Hemphill 2008; Mageria, Makogola & Ndegwa 2008). Vessels have to be licensed and there is an urgent need for a monitoring system (Aloo 2007). This system, which actually has been in the making for some time, should be able to track all vessels operating in Kenyan waters (except artisanal fishers) and to verify the status of licenses and the reported catches. The fleet consists of 'long-liners' and 'purse-seiners' that are required to keep records of their catches but it is likely that much of the catch is not reported (see MRAG 2005).

Long-liners fish with long lines and large hooks. The number of vessels at any time of the year ranges between 20 and 50, depending on season. There is one registered Kenyan vessel among them, the rest are mostly from China and Taiwan. They are required to obtain licenses from the Mombasa Fisheries Office for either one month ($5,000), three months ($7,000) or twelve months ($12,000). In addition, they are required to keep catch records by species, which are registered and accounted for internationally. The Fisheries Office in Mombasa keeps records of long-liner catches; these figures do not appear in Kenyan statistics because catches are taken mostly elsewhere.

Purse-seiners use large nets that close at the bottom. There are 35 vessels of this kind active mostly from Spain and France (operating out of Mauritius). Annual licenses, which are issued by the Director's Office in Nairobi and paid there, cost $20,000 per vessel. No catch records are available and, as far as known, none are kept in Kenya until now.

While offshore fishing is largely the domain of foreign long-distance fleets, the local population is mainly involved in inshore fishing. Artisanal fishing is an important economic activity but depends on the seasons. The *kusi* season is due to south-east monsoon winds, which blow from March to October. High cloud cover, heavy rainfall, river discharge, terrestrial runoffs, cool waters and a deep thermocline characterize this period. Fish catches are lowest during this season as a consequence of fish migrations, decreased fish density and fish activity and reduced fishing effort (McClanahan 1988). The reduced fishing effort results from the inability to fish beyond the lagoon and unwillingness to brave rougher water. The *kaskazi* season has north-east monsoon winds and occurs from September to February. This period offers the best fishing.

Marine fisheries along the coast employed an estimated 10-12,000 fishers in 1999 (as detailed in Chapter 6). Including workers in support industries and household dependents, we estimated that 167-200,000 people, out of a total of 2.49 million living in Coast Province (~7.5% of the population), were wholly or partly dependent on fisheries.[3]

The prospects of the fishing community are negatively affected by the deterioration of coral reefs, the decline in mangroves, the pollution of ocean waters and the existence of Marine Protected Areas. Mangrove forests and coral reefs provide protection to the coastline against the sea, are rich depositories of biodiversity and offer breeding grounds for many marine species (The field guide by Richmond (1997) contains a full review of fauna and flora of the coastal zone in the western Indian Ocean Region). Already, in 1995, it was reported that fishing for the whole Kenyan coast, with 37,000 tons, was near its maximum sustainable yield depending on the fishery. "Artisanal fishing of nearshore reefs are probably beyond Maximum Sustainable Yield (MSY), shrimp, lobster and crab are at MSY while the offshore ... fish activities are probably still below this level" (McClanahan 1996: 54). This implied that fishers basing their survival on fish resources in inshore waters cannot expand to better their future unless, perhaps, they are able to fish deeper waters. The livelihoods of fisher households can also be strengthened by other economic activities so that they are no longer dependent on fishing as the only source of income. How income diversification subsequently affects the exploitation of marine resources is the topic of interest here.

Artisanal fishers themselves can also contribute to the degradation of marine resources, as intensive fishing can affect the ecological balance and result in loss of local biodiversity. Destructive fishing practices, such as the use of

Box 1.2 Two fishing villages

The villages of Takaungu and Uyombo are both situated at the mouth of a large creek and both are protected by coral reefs offshore (Map 2; p. 24). The population in both villages consists mainly of Waswahili, Bajun and Mijikenda but this is where the similarities end.

Takaungu was settled in the early 19th century by members and clients of the Mazrui family. It is likely that the Bajun had already founded a temporary fishing village there before the Mazrui arrived (as they are known to have done at many places along the coast). Certainly other Bajun migrated in numbers to the growing settlement (Koffski 1977) and later Mijikenda also moved to the town. Takaungu has grown considerably in size and where once there were *shambas*, today there are houses and the *shambas* have moved to the outskirts of town.

Uyombo has two parts: an inland village and a landing site. The landing site has a relatively short history with the first settlement dating from 65 to 70 years ago when a Bajun fisher from Lamu decided to build a house and move his family there. Most of the land in or near the landing site is, or was, owned by this family. More people settled but it has remained a small village. Other fishers built temporary shelters where they spent the night when fishing before returning to their homesteads. Many of them were farmers who turned to fishing and whose homesteads and *shambas* were more than an hour's walk away.

The differences between Takaungu and Uyombo were pronounced. Takaungu was considerably larger in size than Uyombo, with many more houses and inhabitants. Uyombo could be reached on foot or by bicycle and although it was possible to reach the area by car, this involved negotiating part of the way through *shambas* and mangrove forests. From Takaungu there was a road connection to the Malindi-Mombasa trunk road and a choice of transport that ranged from *matatu* to private cars, smaller trucks (that supply the shops in Takaungu or carry blocks from the quarry in Timboni) and *boda boda*. Mombasa and Kilifi could be reached within an hour. From Uyombo, however, one had to walk to the trunk road and wait for a *matatu* to Malindi, Watamu or Kilifi.

As a result of its modest size and its poor accessibility, income-generating activities in Uyombo were restricted to either fishing or agriculture, such as fish selling, palm-wine tapping and selling, cash-crop cultivation, plaiting *makuti*, and farm labour. In Takaungu there was a much wider range of income possibilities such as furniture making, block cutting, building construction, teaching and so on. In addition there were shops and small eating places. There were *fundis* and tailors resident in Takaungu but not in Uyombo.

Source: Versleijen (2001)

small-mesh nets, beach seines, poison and explosives (Ochiewo 2004), can alter the terrain as well as the ecological balance of the reef and seafloor (Mangi & Roberts 2006). Local fishers generally do not approve of destructive fishing methods since they are aware that these will ultimately lead to poorer catches. Indeed, nearly all fishers were concerned with the degradation of marine resources and declining fish catches (Hoorweg *et al.* 2006). Among the reasons given for these trends were the increasing number of fishers, gazettement of no-take-areas, rough weather and competing fisheries such as commercial trawling.

Fisher households can continue to draw a livelihood from fishing with access to better fishing techniques, enough desirable species in catches and proper marketing facilities. This requires sustainable fishing methods in combination with improved care of breeding grounds to assure the long-term future of the fisheries. However, an increase in the use of illegal and destructive fishing methods is equally possible. For example, there have been reports of the placing of traps in breeding sites, the use of poison in Ungwana Bay[4] and even the occasional use of explosives on the south coast (East African 2000). (Dynamite is commonly used in Tanzania; see Guard & Masaiganah 1997; Horrill & Makoloweka 1998; Jacquet & Zeller 2007). Although the sales of shells and corals are banned in Kenya, they are still being collected. It is also likely that local aversion to Marine Protected Areas will increase. Resistance was already expressed to the proposed Diani Reserve, which then was rejected by the local population (Alidina 2005). Nevertheless, the Reserve was officially established by the authorities in 1995 (WIOMSA 2007).[5]

Whatever happens, fisher households, out of necessity, will have to enlarge their resource base with other economic activities if they have not done so already. Opportunities for maritime employment are few. In Malindi Marine Park, Bajun fishers were given permits to operate glass-bottom boats to take visitors for goggling on the reef. Some fishers found employment as crewmembers on sport fishing boats. Possibilities for non-maritime employment depend on the existing opportunities and the economic footholds that households already have in the local economy, such as farming and cottage industries. Households with non-maritime employment strengthen their livelihood strategies and improve their household security. These fishers become less dependent on fishing and it is expected that they will put less pressure on marine resources and develop more positive attitudes towards conservation measures.

In spite of the impending plight of fishers, little is being done. Fishers have been largely neglected and few, if any, alternative forms of livelihood are available to them. Furthermore, there is little knowledge about social and economic characteristics of inshore fishing. Income opportunities of fisher households differ greatly as they depend not only on the characteristics of the coastline and the fishing grounds but also on other geographical as well as social and cultural factors (see Box 1.2). The impression is that household incomes and income composition vary greatly among fishing villages and within villages. In some parts of the coast, fishers are regarded as the 'poorest of the poor'; elsewhere they are considered 'well off' (Mwadime 1996). Moreover, little is known about other resources that fishers may possess, the nature of these resources, and to what extent households are dependent on them.

Research in the social and economic conditions of fishers is needed to understand their responses to the deteriorating situation. Firstly, this is important for the future of this group which has thus far received little political or research attention. In general, smallholder households in coastal Kenya try to diversify their incomes with cultivation of food crops for home consumption, income from cash crops, livestock and non-farm employment. Income diversification is an important factor in food security and household livelihoods (Hoorweg, Foeken & Klaver 1995). Fishers do not easily abandon the family profession in which they have been raised and for the moment, therefore, income diversification of fishers appears the most suitable strategy. Secondly, this is important for the protection of the marine environment, since fishers can potentially cause extensive damage. If fishers can secure economic alternatives they may become more prudent in their fishing practices and, hopefully, may even become guardians and stockholders of the maritime heritage. Information and understanding of the relation between household strategies of local fishers and resource management is vitally important.

Outline

Between 1999 and 2001, a team of researchers and students studied household characteristics and resource conservation among artisanal fishers. *Resource Di-*

versification and Management among Coastal Fisher-folk in Kenya was a joint project of Moi University (Kenya), Ben Gurion University (Israel) and African Studies Centre (Netherlands). In all, the project was comprised of four main surveys and four support studies. The surveys covered the characteristics of fishers and fishing on this coast, fish catches, trading and marketing of fish and livelihoods of fisher households. The support studies were on catch composition and reproductive biology of fish, fish sales and marketing, income diversification of households and resource conservation by fishers. Research methods are detailed in Chapter 2.

The main focus of the research was on income diversification of fishers on the Kenyan Coast, the pressure on marine resources and the relation between the two. For income diversification, the attention was mainly on two questions, namely, how incomes of fishers compared with non-fishers and how diversification affected the incomes of fishers and the incidence of poverty among them. For pressure on marine resources, the attention was on number of fishers, access to fishing grounds, type of gear and fishing intensity. And, finally, we examined whether there was a relation between income diversification of fishers and pressure on marine resources, that is, whether fishers with income from more than one source ultimately exacted less pressure on the marine environment.

This monograph brings together the findings of the surveys and support studies. The choice was between a comprehensive compilation of all results or a smaller monograph with the essential findings. The latter was considered the preferred option and with fewer than 150 pages we hope that we have managed to be both succinct and comprehensible. The narrative follows the order of the four surveys: general characteristics of fishers and fishing (Chapter 3), fish landings and fish traders (Chapter 4) and fisher households and their livelihoods (Chapter 5). The pressure on marine environment and the relation to income diversification of fishers are discussed next (Chapter 6) by combining results of the fisher survey, the household survey and the relevant findings of the support studies. Findings of the support studies have otherwise been presented in boxes that accompany the main text.

Map 2 Malindi and Kilifi coast with coastal tracts and landing sites covered in the fisher survey and trader survey

2

Talking to fishers

Study area

The East African coast stretches for 5,500 km along the Indian Ocean and includes the coastlines of Somalia, Kenya, Tanzania and Mozambique. The Kenyan coast covers about 600 km from the Somali border in the north to the Tanzanian border in the south. The southern part of the Kenyan Coast, below Malindi, consists of tiers of pleistocene reefs above and below sea level. North of Malindi, the coast is formed by broad sedimentary plains drained by the Tana and Athi-Sabaki rivers. These rivers dominate the coastline due to the sedimentation they bring from the agricultural and industrial hinterland. The continental shelf is narrow except off Malindi and the Tana River mouth (Frazier 1993; UNEP 1998a).

The Coastal Region is generally low and is characterized by the extensive fossil reef, which lies a few meters above the present sea level. The coastal plain is backed in the interior by a line of hills that rarely exceed an altitude of 300 m except in the southern parts where the Shimba Hills reach around 1,000 m above sea level. Further inland, the Taita Hills rise to an elevation of 1,500 m (Foeken 2000). Most of the shoreline, apart from the Malindi area, is receding as a result of coastal erosion. Sand supplies from rivers and coral reefs are not sufficient to keep up with the rise in sea level, and the problem is further exacerbated by coastal development.

The climate on the Kenyan coast is dominated by large-scale pressure systems from the Western Indian Ocean and the two distinct monsoon periods. The north-east monsoon prevails from November to March, the south-east monsoon from May to October. The tidal range is about four kilometres. Annual rainfall on land close to the sea is about 1,000 mm whereas a few kilometres inland it is only 700 mm. The shoreline exists of rocky fossil coral cliffs, mangrove stands and sandy beaches (Obura 2001; UNEP 1998a).

A fringing reef parallels the shores to the south and to the north anywhere from 0.1 to 1.0 km offshore interrupted at creek and river outlets (Map 1, p. xii). Coral reefs are among the most productive ecosystems with great biodiversity (McClanahan & Obura 1996). The reefs also contain special habitats like tidal pools and, with shorelines, are important fishing grounds for the artisanal fishers. Other marine resources include sea grasses and seaweeds, mangroves, sea turtles, marine mammals, crustaceans and various billfish (Aloo 2000). Special features include the Lamu Archipelago with its extensive mangrove forests; the Tana River which is Kenya's largest river and discharges through a complex wetland system into Ungwana Bay; the Sabaki River which incorporates the Athi and Galana Rivers and discharges north of Malindi; Kilifi Creek; the coral islands like Wasini Island, Chale Island and Funzi Island, and Gazi and Funzi Bay.

Coast Province has six administrative districts that border the sea, namely from north to south: Lamu, Tana River, Malindi, Kilifi, Mombasa and Kwale (Map 1). The study area is situated in Malindi and Kilifi districts, extending from Ras Ngomeni (the Ngomeni peninsula) to Takaungu Creek, a coastline of roughly 125 km and consists of five coastal tracts with two landing sites each. From North to South these are the Ngomeni, Malindi, Mida, Kilifi[6] and Takaungu coastal tracts. This more or less covers the coast of Kilifi and Malindi districts excepting about 25 km north and 25 km south. This choice was made for logistic reasons and to ensure cultural consistency.

In 1999, the total population of Coast Province was 2.49 million people and that of Malindi and Kilifi districts was 825,855 people (Kenya 2002a; Statoids 2008). The coastal inhabitants belong to various ethnic groups, mainly Mijikenda, Arab and Swahili. They are generally poor and rely heavily on simple, traditional methods of food production, which include subsistence farming and artisanal fishing. Fishing provides the principle source of animal protein and is an income earning substitute, especially during the high fishing season. The predominant nearshore activities include artisanal fishing, shrimp trawling, other commercial extraction and tourism, but marine resource use is largely unregulated (Obura 2000).

The *Ngomeni* coastal tract is characterized by open access to the sea, mangroves, mud flats and sandy beaches but the absence of a fringing reef. There are two landing sites, Ngomeni village and opposite Robinson's Island. The *Malindi* coastal tract is near the Malindi National Reserve with one landing site in Malindi town at the very end of the reef, which at this point is polluted and covered by sediments from the Sabaki River. The second landing site, Mayungu, is a small cove amidst dry, rocky land in the middle of the Malindi Reserve. The *Mida* coastal tract consists of the Watamu and Uyombo landing sites. Watamu is situated within the Marine Park of that name; Uyombo is adjacent to this Park but fishers have to pass through the Park to reach their fishing grounds. The first site is situated on a sandy beach with nearby coral rocks towering over the sea and pounding waves. Uyombo, the second site, lies at the entrance of Mida creek, a large inlet that falls largely dry during ebb tide. The *Kilifi* coastal tract consists of the landing sites of Bofa and Kilifi Ferry that are both within easy reach of Kilifi town. Bofa is further up the coast with small rocky outcroppings, while Kilifi Ferry is situated at the mouth of the deep Kilifi creek that serves as a harbour for large vessels and pleasure yachts. The *Takaungu* coastal tract is characterized by coral soils and palm cover and consists of landing sites at Takaungu town and Shariani. The coral reefs here are patchy in nature and further out to sea. Takaungu town is situated at a deep creek that falls largely dry at ebb tide: Shariani is on the seaside and is steep and rocky.

Study design

During the identification and design stage of the project, April-May, 1999, all official and unofficial landing sites in Kilifi and Malindi District were mapped and essential site-information recorded, for example, number and types of boats; fisher residence; fisher ethnicity; public services; road access; cooler facilities; number and types of traders; gender of traders; other fisher-related economic activities; period of peak activity; and other unique characteristics.

The research consisted of surveys and supporting studies. Four independent surveys reviewed artisanal fishers, fish landings, fish traders and fisher households respectively. The fisher and trader surveys covered five coastal tracts with ten landing sites (Map 2, p. 24)*:

*Ngomeni CT; *Malindi CT; *Mida CT; *Kilifi CT and *Takaungu CT.

The survey on fish landings and the household survey were restricted to 4 selected landing sites that differed in nearness to Marine Protected Areas, where fishing conditions were presumably better, and that also differed in access to employment opportunities (Map 3, p. 34):

*Ngomeni LS; *Mayungu LS; *Uyombo LS and *Takaungu LS.

Sampling generally focused on fishers at the landing beaches or started from there and continued to households or markets. The respective surveys and support studies are detailed below.[7]

Survey of artisanal fishers

The 'Fisher Survey' was carried out between June and October, 1999, and covered 5 tracts of coastline, each represented by 2 landing sites (Map 2). At each landing site 20 fishers were randomly selected and interviewed, either on-site or at their homes, resulting in 40 fishers for each coastal tract. They were interviewed by one of the research assistants in the vernacular. The following information was covered: type and frequency of fishing activities; standard catch data; crew and ownership arrangements; socio-economic and household characteristics; and catch utilization (subsistence/sales characteristics). In total, 199 interviews were recorded.[8]

The sample consisted of 95 boat captains, 81 crewmembers and 23 independent fishers. Most fishers lived within a distance of three km from the landing

---

* Some coastal tracts and landing sites share the same name. Where this may lead to confusion, the notation CT (Coastal Tract) and LS (Landing Site) are added to the names concerned (For example, Takaungu CT which comprises Takaungu LS and Shariani LS).

site (86%) and most were younger than 39 years of age (64%). Over 50% of the fishers in the sample had fished for more than ten years. Many had never received formal education (44%), particularly the older fishers, although there were also younger fishers who had dropped out of school. There were only small differences among the coastal tracts in respect to age and education of the fishers. The majority of fishers (59%) listed a 'single' economic activity, that is, they did not engage in any other employment than fishing, not even farming. Consequently, 'multiple' activities, meaning that fishers listed other employment next to fishing, were reported by less than half the fishers in the sample (41%).

Survey of fish landings

The survey of fish landings is herein referred to as 'Catch Survey'. Fish catch data were recorded between May, 1999, and March, 2001, at four landing sites (Map 3): Ngomeni LS, Mayungu LS, Uyombo LS and Takaungu LS (In Takaungu, data collection was till April, 2000). They were recorded by local assistants, twice weekly, on random days for all fishers who brought in catches on these days. Essential information was recorded including vessel, crew, gear used, fishing grounds, fish catch, fish species and income. By the end of March, 2001, 8,164 records had been compiled during 611 observation days at the four sites.

The assembled data were treated in five steps as follows:

> Some crew sizes were extremely large – up to 35 members. This occurred, for example, in groups of divers that were commercially organized and that were taken by motorboats to the grounds. The groups included other exceptional forms of fishing as well. Observations on the landings of large crews of 6 members or more were excluded (191 records).

> In Ngomeni, Mayungu and Uyombo, the starting months of May-June, 1999, and the end months of January-March, 2001, had low numbers of observation days. These data were omitted from the analysis (458 records).

> As a result, observations in Ngomeni, Mayungu, Uyombo covered 18 months with the months of July-December represented twice. To arrive at a

representative yearly estimate, the latter months were weighted with a factor of 0.5.

> In Takaungu, observations covered exactly one year but the number of observation days were low in May, 1999, and April, 2000. The latter observations were weighted with factors of 4.5 and 2.25 respectively.

Table 2.1 lists the numbers of the remaining 'catch records' collected in each of the four landing sites together with the resulting weighted numbers.

Table 2.1 Number of catch records by landing site

	Ngomeni LS	Mayungu LS	Uyombo LS	Takaungu LS	Total
Raw number	1785	2234	2593	903	7515
Weighted number	1228	1520	1751	1011	5510

Survey of fish traders

The 'Trader Survey' covered the same coastal tracts as the Fisher Survey described above and the same landing sites (Map 2). This survey was done between December, 1999, and March, 2000. At each landing site, traders were randomly selected and interviewed by one of the research assistants either onsite or at their homes. In total, 186 traders were interviewed at the five coastal tracts: Ngomeni (N=32), Malindi (N=42), Mida (N=37), Kilifi (N=43) and Takaungu (N=32). The following information was covered: buying and selling data (prices, volumes, composition), storage and transport, destination as well as socio-economic household characteristics.

Traders came from the landing sites or nearby villages (27.5%), the nearest villages on the tarmac road (16.1%) and urban centres in the area (36.6%). Most traders were in the age groups of 20-29 years (42.1%) and 30-39 years (27.3%). A large number of traders had not attended any formal education (37.0%). There were only small differences among coastal tracts in respect to age and education of the traders. The majority of traders were of Mijikenda extraction (69.4) and far fewer were of Bajun origin (22.6%). There were almost no traders of Bajun origin in the Kilifi and Takaungu coastal tracts.

Survey of fisher households

The 'Household Survey' covered four landing sites that differed in proximity to Marine Protected Areas and potential access to employment in nearby urban centres (Map 3). Two sites were situated near a Marine Protected Area, one with employment opportunities in the vicinity (Mayungu LS) and one without (Uyombo LS). Two sites were not situated near protected areas, one with (Takaungu LS) and one without employment opportunities nearby (Ngomeni LS).

This survey included both fishers and non-fishers. The selection started with the identification of fishers who were seen regularly during the Catch Survey.[9] This group included boat captains and independent fishers who operated alone but with the common characteristic that fishing was an 'entrepreneurial' activity and that they were responsible for their craft and gear and the risk of loss or damage. The selected fishers were met at the beach and a visit to their homes was arranged where they were interviewed. All fishers were asked to identify non-fishers; that is, the nearest neighbours where the head of the household was not a fisher. Each boat captain was also asked to identify the household of a regular crewmember living nearby who, in principle, received payment or a fixed share of the catch. Next, these groups of households were contacted and also interviewed. Since the pairs of fishers and non-fishers as well as the pairs of boat captains and crewmembers were living in the same vicinity, we assumed that external variables such as farming conditions and (distance to) employment opportunities were matched.

Data were collected from October, 2000 to March, 2001, on the following areas: living conditions, household composition, employment characteristics, farming activities, fishing activities, incomes of the head and other household members, resource conservation and food consumption. In all households, the head of the household and his wife were interviewed (and the fisher concerned if he was not head of the household) by one of the research assistants in the vernacular.

A total of 213 households were visited and interviewed (83 boat captains and independent fishers[10] who are grouped together for purposes of analysis and who are referred to as boat captains in short, 50 crewmembers, and 80 non-fisher households). Heads of households were mostly younger than 40 years of age (55.3%) and a third had not attended any formal education (31.4%). Three-quarters of the heads of households were married. The average household size was 7.7 people among fishers and 6.8 people among non-fishers. In respect to material conditions, the quality of housing and hygiene was slightly better among the non-fishers.

In 22 cases, the fishers and crewmembers were not heads of households but were other adults, for example, grown-up sons, living with their parents. These households were included in the initial analysis of household economy and household income. However, they were omitted from later analysis of income diversification, together with the group of non-fishers, for reasons of definition. The remaining sample numbered 111 fisher households.

Supporting studies

Apart from the surveys listed above, four supporting studies examined certain aspects in more detail, namely catch composition and fish biology, fish trade and marketing, livelihood strategies of fishers, and traditional ways of marine conservation. The studies have been written up independently as M.A. theses. Short descriptions of the methods of these studies are presented below. The findings have been used in the text where relevant and are also presented in separate boxes.

Fish Catch Composition and Some Aspects of Reproductive Biology of Siganus sutor *along the Malindi-Kilifi Marine Inshore Waters (Mohammed 2002).* Data on catch composition of artisanal fishers were recorded twice a week for a period of six months at four landing sites: Ngomeni LS, Mayungu LS, Uyombo LS and Takaungu LS. The specimens were identified up to species level using relevant field guides (FAO 1985, Smith & Heemstra 1991; Richmond 1997). Samples of rabbitfish from *malema* traps at Mayungu LS were collected twice a month for a period of four months (November-February) and examined for total length of snout to end of caudal fin (TL; cm), standard length from snout to beginning of caudal fin (SL; cm), weight (g), gonad weight (g), sex and fecundity. A fish measuring board was used in the field to measure the fish to the nearest 1cm and a top-loading balance was used to weigh individual fish to the nearest

1g. Gonads were removed, placed into vials and stored on ice and, in the laboratory, they were weighed to the nearest 0.1 g using an analytical balance. The ovarian maturation cycle was determined using histological techniques and size at first maturity was established.

Processing and Marketing of Fish among the Coastal Fisher-Folk (Wamukota 2002). Fish traders at four landing sites were surveyed between October, 2000, and March, 2001 (Ngomeni LS, Mayungu LS, Uyombo LS and Takaungu LS). Market centres were selected in close proximity to the four landing sites (Ngomeni, Malindi, Matsangoni and Takaungu). Traders from all the landing sites and market centres were sampled. Data collection techniques included structured questionnaires, interviews, observation, and the use of secondary data from various sources.

An Empty Sufuria: The Effects of a Marine National Park on the Livelihood Strategies and Income Diversification of Fishermen Households at the Kenya Coast (Versleijen 2001). This study focused on the livelihood strategies of artisanal fishers and their attitudes towards resource conservation, indigenous conservation practices and the presence of a Marine Protected Area. Information was collected through questionnaires, participant observation, life and career histories, network analysis and genealogies in Uyombo LS and Takaungu LS, with additional information from fishers in Watamu LS. Discussions were held with the fishers, Kenya Wildlife Service employees and employees at the Watamu Marine National Park. The period of study was June-October, 2000, and included 21 respondents from Takaungu and 23 from Uyombo.

Reef Fisheries in Kilifi and Lamu Districts: Fishing Practices, Awareness of Resource Degradation and Traditional Ways of Conservation among Artisanal Fishermen (Tunje 2000). Fishing methods, reasons for their choice, and their impact on coral reefs were investigated. Indigenous environmental conservation efforts, fishers' alternative sources of income, and attitudes towards environmental conservation were also studied. Data were collected from August, 1998 to February, 1999. Main study sites were Mayungu LS and Takaungu LS (25 fishers interviewed in each site); additional information was collected in Uyombo LS and Mbuyuni LS and sites in Lamu District. Methods included formal questionnaires, in-depth interviews, informal discussions and participant observation.

Map 3 Malindi and Kilifi coast with landing sites covered in the catch survey and household survey

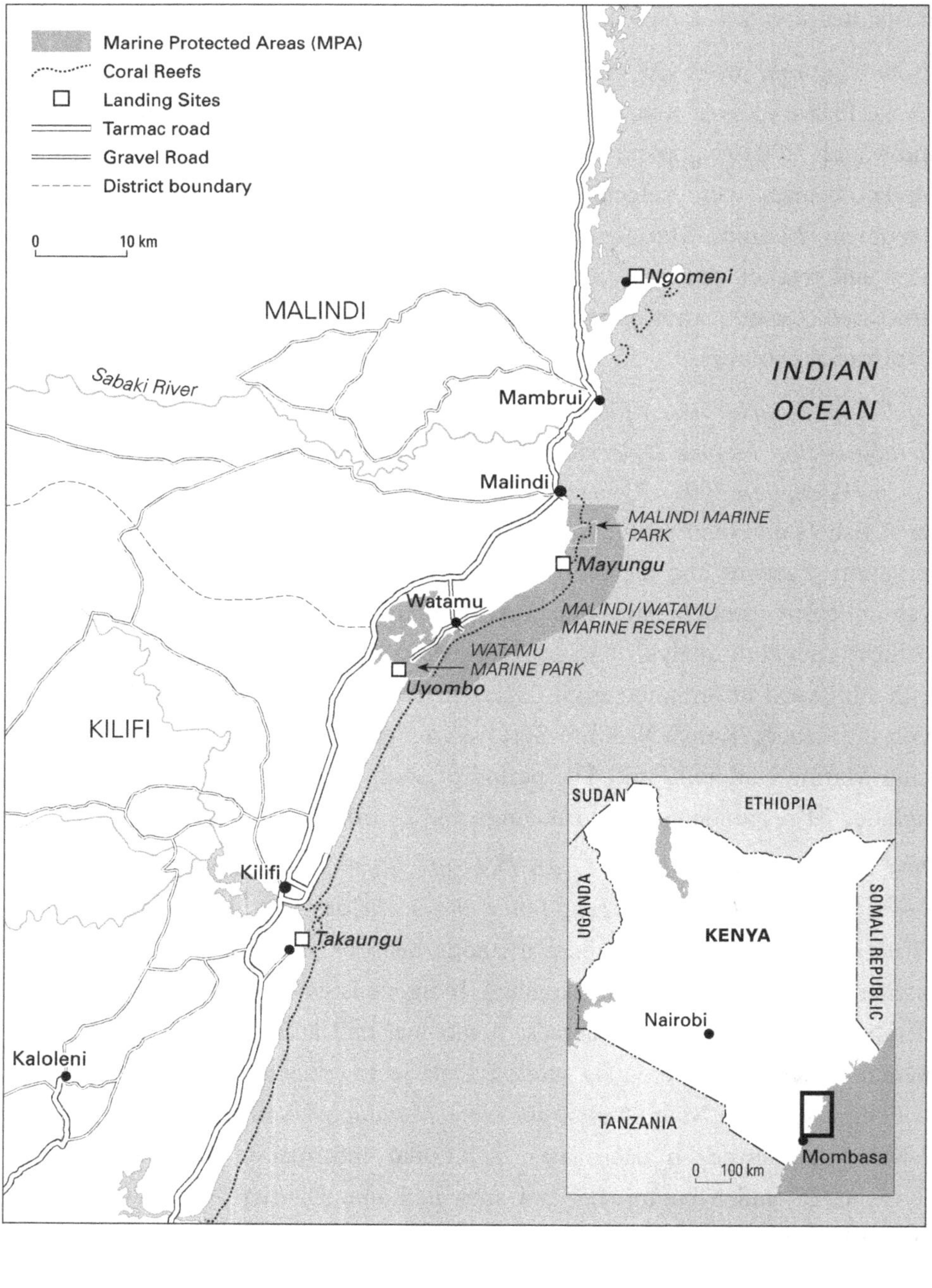

3

Artisanal fishers and their craft*

Artisanal fishers comprise boat captains, independent fishers and crewmembers using traditional gear, lines and nets as well as fishers who rely on diving or collecting. The fisher may use the craft or gear himself or hire it out to other fishers whose payment contributes directly to the livelihood of the fisher. In this respect, crewmembers are also considered as fishers. Handling of the landed catch is done by fish traders who either buy to sell or process before sale. Fishers are involved in the actual catching of fish. Also very closely associated with the fishers' activities are boat-builders and repairers, gear producers, firewood providers and icemakers.

It is difficult to establish the number of people directly involved in fishing because of the seasonality of the occupation. Many fishers try to evade paying license fees, thus failing to appear in government records. This creates serious weaknesses in the estimates used in policy planning for management of small-scale fisheries. The first survey of the number of fishers was done in 1948 by the Assistant Fish Warden who visited the major fishing villages along the coast and reported 1,019 boats and over 2,500 fishers (Kenya 1950b). In 1961-1963, it was estimated that the number had increased to 5,400 fishers and 1,700 ves-

---

* The findings in this chapter are from the Fisher Survey at five coastal tracts, unless indicated otherwise.

sels (FAO 1966a). In the late seventies, there were 2,000 boats with 6,000 fishers according to Okidi (1979). FAO (1984) gave an estimated number of 1,800 fishing vessels, while Ardill & Sanders (1991) reported a figure of 6,250 fishers and Wamukoya *et al.* (1996) mentioned 5,000 officially registered fishers with 2,000 boats. Glaesel (1997a) estimated 6,000 fishers but she also mentioned a much higher estimate of 4,000 boats and 12,000 fishers by the Ministry of Tourism and Wildlife. According to the Department of Fisheries (1996-1998) there were about 1,000 small-scale fishers on the coast of Kilifi and Malindi District but the number is likely to have been much higher according to rough estimates of the fishers themselves which arrive at about 2,500 fishers for this part of the coast (see Chapter 6).

The fishers were organized in fishing communities headed by a 'chairman' who acted as a link to the government fisheries authorities. It has to be kept in mind that fishers are not really a homogeneous group. The majority of fishers targeted finfish but they used different types of vessels. Others fished for crabs, octopus, and lobster. Some fishers used nets, others traditional traps, spear guns, hand lines or fixed fences. This diversity in craft and gear implied a high disparity in interests and stakes. Since there were no co-management arrangements existent at the time, enforcement of fishing regulations was ensured by sporadic patrols by the District Fisheries Officer. Those who breached regulations were fined or suspended for varying periods or in some cases the authorities simply confiscated and destroyed the offender's gear.

The active fisher population comprised of boat captains, independent fishers and crewmembers. Boat captains and independent fishers fished as an 'enterprise' in that they were responsible for their craft and gear and the risk of loss or damage. Boat captains employed fellow fishers as crewmembers. The crew size including the captain was, on average, 3.6 persons per boat.[11] Nearly all active fishers are men but isolated cases have been reported of women participating in fish capture, notably octopus, prawns and shells (Malleret-King *et al.* 2003).

Vessels were owned by owner-captains (45%) or hired from a *tajiri* (almost 40%). A *tajiri* is a trader and entrepreneur who buys the catch that fishers bring in and also leases vessels and gear to fishers in return for a share of 20-50% of the daily catch; the remainder is shared among the fishers operating the vessel (Glaesel 1997a). Maintenance of the fishing equipment was the responsibility of the fishers. There were, however, other ways of sharing the proceeds, which usually resulted in greater benefit to the trader. For instance, the *tajiri* sustained the fishers by providing soft loans in times of financial hardships, making them highly dependent on him. This system was cited by fishers as one of the major factors in their economic backwardness. However, the *tajiri* may have had difficulty in collecting his share because the fishers had ways of hiding the true size and composition of the catch. In addition, sometimes boats were hired from fellow fishers (15%).

The gear usually belonged to the captain but was provided by a *tajiri* in 10% of the cases or by another crewmember in even fewer cases, but there were differences among landing sites. In the coastal tracts of Malindi and Mida, the role of the *tajiri* was rather small. In Malindi the diminished role had to do with the presence of a co-operative, which provided financial support to the fishers through a loan system. For Mida, it was probably due to the modest catches there. In Ngomeni, Kilifi and Takaungu the *tajiri* played a more important role, particularly in Ngomeni where in 20% of the cases he supplied the gear. This probably was because of the larger vessels and heavier gear used in fishing at this part of the coast, which were more costly.

One-third of the fishers always landed their catch at the same site. Two-thirds of the fishers visited other landing sites on the occasion when they fished elsewhere or because they deemed marketing opportunities better there. Almost half the fishers in Malindi and Mida (the tracts with the lowest catches; Chapter 4) mentioned that they frequented three different sites, which suggests that they moved more often along the coast in search of better fishing grounds.

The catch was usually divided as follows: the owner of the boat typically took 40-50% of the catch, either the *tajiri* or the captain himself; the remaining half of the catch was divided in equal shares among the captain and crewmembers with an extra share for the owner of the gear. Sometimes, an experienced captain was entitled to an extra share when taking out an inexperienced crew. These arrangements, however, were flexible and subject to change.

There were no fishers who fished only for subsistence needs. Fishers usually sold most of the catch and took some fish home for consumption. The part of the catch that is taken home, locally referred to as *kitoweo*, has been estimated to account for 4% of the artisanal catches world-wide (UNEP 1998b). Almost half the fishers kept some fish as bait for the next day. Two-thirds of the fishers sold the fish as is, in fresh condition. The percentage of fishers doing this was highest at Malindi and Kilifi, sites that had ready access to markets, followed by Mida and Takaungu. At Ngomeni, the most remote tract, two-thirds of the fish were gutted and dried.

Obstacles mentioned by fishers ranged from low catches to cheating by traders. Lack of equipment was most often mentioned (89%) as a constraint because of the high costs involved in replacing worn-out, damaged or lost nets. However, equipment associated problems were not mentioned as often in Ngomeni as elsewhere. Next in the list of obstacles were financial problems (58%) which were either related to low incomes or lack of money for equipment followed by transport (37%) and marketing problems (30%). There existed, however, considerable local differences. Fishers in Malindi and Kilifi almost exclusively mentioned equipment and financial problems whereas in Ngomeni, transport and marketing were major problems. In Mida, transport was a bottleneck, which was not the case in Takaungu because *matatus* and buses reached there. In addition, there were site-specific circumstances, notably the competition by trawlers that approached close to the coast in contravention of existing regulations (Fulanda 2003). In Mida (Tunje & Hoorweg 2001), there were many complaints about the presence of the nearby Marine Park, which occupied good fishing grounds that were off-limits and also made the fishers subject to inspection by the game wardens, to examine whether they had been fishing illegally in the Park. Cinner & McClanahan (2006) also reported that many resource users had a negative perception of marine parks. The relations between the local fishers and Kenya Wildlife Service (KWS) were tense. Fishers claimed that they were sometimes arrested on their way to the Reserve and then accused of fishing in prohibited waters. They even complained about beatings and fishing gear and vessels being confiscated (see also Box 6.1, p. 94).

Finally, the number of fishers who were members of a co-operative was quite low, only 20%. The percentage was higher in Malindi and Takaungu, although even here no more than half the fishers were organized. Of the ones who were, three-quarters complained about lack of benefits from their membership and only a quarter (less than 10% of all fishers) mentioned loans for purchase of gear and improvement of marketing facilities as benefits from the co-operatives.

Fishing vessels

Fishing or catching fish requires the use of craft to reach good fishing grounds and use of gear such as traps, nets or hooks. A large variety of fishing vessels and gear are used by artisanal fishers all over the world. In all cases, the choice of gear depends on the target species just as the choice of vessel depends on the safety conditions of the waters.

Artisanal fishers in the study area used six types of vessels known as *mtumbwi, hori, ngalawa, dau, mashua* and *jahazi*. They are described in detail in Table 3.4 (p. 49), which gives the length of the vessel, the clearance, manner of construction, propulsion, steering mechanism and rudder fixture if any, necessary crew and buying price. Glaesel (1997b) published a detailed description of fisher vocabulary in respect to vessels, gear and materials used for the construction of fishing equipment. She collected the terms mostly in the Mombasa-Mtwapa area; however, fishing vocabulary shows considerable local variations, as was pointed out by the same author. The terms below are for the Malindi-Kilifi area.

A *mtumbwi* is a dug-out canoe of about 4 m in length with a curved bottom and made from a tree trunk (a canoe made of planks is called *hori*). It is propelled using an oar and is mainly used by gill net and hand line fishers inside the reef. Canoes are sometimes fitted with outriggers and small sails called *ngalawa*, which are few and not built locally but on the island of Pemba. *A dau* is built from plankwood, has a flat bottom and is usually propelled by sail (*tanga*).[12] It has an average length of about five metres and is spacious, making it suitable for *malema* (trap) fishers inside the reef. *Mashua* are fishing vessels used mainly for out-of-reef fishing. They are larger in size, about ten metres in length, made from plank wood and have sails. Night fishers who use long lines and floating nets prefer them. A *jahazi* is an even larger vessel, often used to transport cargo, but also used for fishing. Usually, only *mashua, jahazi* and motorboats reach the open sea, restricting the smaller vessels to the inshore waters. However, smaller vessels can also venture into the deep waters if the weather is suitable.

Box 3.1 Fishing practices and their environmental impact

About half the fishers fished once a day; one trip taking about four hours. They generally fished for 6 days in a week and rested one day. However, in areas where fishing was poor, many fishers were compelled to fish twice a day, or engage in day and night fishing. In general, frequency of fishing depended on resources and accessibility, the type and size of the fishing vessel, and the distance the fisher had to cover. The number of fishing trips per day was higher among low-income fishers.

Traditional gear such as fish traps and fish fences were considered less damaging but their use was on the decline. Spear guns were prohibited in Marine Protected Areas, although some fishers used them at night. Spear fishing can be destructive as a consequence of fishers stepping on corals when snorkelling and of arrows missing their targets and striking corals. Spear guns were often used in combination with metallic rods which were used to break and (or) overturn the corals where fish seek refuge. However, when used with care, spear guns can be sustainable in exploiting coral reef fishery resources and help with selective exploitation. Traditional poisons and explosives, also illegal, were used sometimes by fishers who could not afford fishing gear. There was some use of fish poison to the north of the Malindi coastline which indiscriminately killed other organisms apart from fish species. The northern part of Malindi district was remote and, hence, regular patrols by the Fisheries Officers difficult.

Modern, manufactured gear, that is nylon nets and lines were preferred by most fishers. Nets were most common with preference for gill nets with mesh sizes between 3.0 and 6.5 inches, which caught only large and mature fish, leaving small juveniles to pass through. According to the Fisheries Act (Kenya 1991), it was illegal to use a net of mesh size less than 1.95 inches (50 mm, measured diagonally), unless used for sardines, which grow to approximately 2 inches. The use of beach-seines, usually very long nets (to 100 m) with very small mesh sizes, was common in Lamu. These nets were dragged along the seabed harvesting targeted but also many non-targeted fish species. They destroyed corals which are fish breeding, feeding and spawning grounds. By-catch of non-targeted fish species and immature fish in trawl-nets can comprise more than 70% of the catch (Fulanda 2003).

The choice of gear was determined mainly by the fisher's knowledge and experience. Also, fishers preferred gear which resulted in relatively high catches. The price of the gear was another factor that influenced choice. Because of regular patrols by KWS officers in the protected areas, fishers used mainly recommended gear, particularly gill nets. This was in contrast to the unprotected areas where fishers rarely considered the environmental impacts of their gear and used many types, including small mesh nets.

Source: Tunje (2000).

Canoes and outrigger canoes accounted for a third of the fishing craft and the *dau* for another third, which means that about 70% of the craft were meant for reef and in-reef fishing (Table 3.1). *Mashua* and *jahazi* accounted for some 18% of the fishing crafts and less than 10% of the crafts were motorised. Lack of suitable vessels dictated that most of the fishing efforts were concentrated inside the reef and far less often in the deep waters. There were quite large differences in this respect among the coastal tracts, depending on the local marine environment.

The number of fishers on board depended on the size and type of boat. Small canoes carried one or two fishers, while big canoes could accommodate three fishers. The outrigger canoe and *dau* also carried more than two fishers while *mashua* and motorboats carried a maximum of six fishers.

Table 3.1 Fishing vessels by coastal tract (%)

	CT Ngomeni (N=40)	CT Malindi (N=39)	CT Mida (N=39)	CT Kilifi (N=40)	CT Takaungu (N=40)	Total (N=198)
Canoe	22.5	25.6	17.9	57.5	50.0	34.8
Dau	5.0	61.5	56.4	27.5	22.5	34.3
Mashua	42.5	10.3	7.7	–	5.0	13.1
Jahazi	22.5	–	2.6	–	–	5.1
Motorboat	–	2.6	5.1	12.5	17.5	7.6
Other	7.5	–	10.3	2.5	5.0	5.1
	100	100	100	100	100	100

Source: Fisher survey

Fishing gear

Fishing gear were either locally made or manufactured industrially. Table 3.5 (p. 50) provides specifications of overall size, mesh size, deployment, day or night use, minimum crew needed, type of boat required, buying price and costs of extras needed. The traditional gear were usually homemade from locally obtained materials and were generally cheap to make. They included portable traps (*malema*), the fixed fence (*uzio*), spear guns (*bunduki*) and poison (*mchupa*). *Malema* were made of wood and reed strips, which were woven into

Box 3.2 Fisher awareness of resource degradation and traditional conservation

Fishers who operated adjacent to and in the Marine Reserves were generally aware of the fishing prohibitions. In addition, they were more environmental sensitive than those in unprotected areas and they observed the fisheries regulations more often. This was in contrast to the unprotected areas where fishers faced less enforcement and were less sensitized to regulations. The fact that fishers claimed not to fish in the Marine Parks also confirmed awareness of the no-fishing zones. Nevertheless, some fishers were prepared to contravene regulations and poached in the Parks.

Fishers complained that they did not receive any financial benefits despite the large sums of money that the Parks generated. Various ways were suggested by the fishers on how they could benefit from the Parks, including community projects, loans for fishers and opening of Parks for fishing during the *kusi* season. Some even argued for the dissolution of the Parks.

Fishers in the Reserves used mainly recommended gear. Nets with small mesh sizes that caught small immature fish were used mostly in unprotected areas. The large majority of fishers considered beach-seines to be the most environmentally destructive; harmful to the coral reef and fish juveniles. Gill nets were perceived as environment friendly.

Most fishers appreciated the importance of marine environmental conservation and were prepared to participate in projects aimed at conserving fishery resources if they were promised incentives to improve their incomes. Fishers generally expressed willingness to reduce fishing frequency and to consider alternative sources of income. The few fishers who showed themselves unwilling were usually uncertain about the future of the promised incentives. Some argued that they had received promises in the past but nothing had materialized. Some fishers said that there was nothing else they could do as they had been fishing since their childhood.

Half the fishers observed certain taboos. The taboos did not deal directly with the marine environment but mostly with personal safety at work, cleanliness and hygiene, and good fish handling practices. There was no mention of indigenous conservation practices with one exception, the *sadaka*, a traditional ceremony with fishers offering sacrifices and saying prayers to the ancestral spirits in a 'sacred' place (*mzimu*) next to the sea. However, local fishers have largely lost respect for this ceremony and fish regularly in areas adjacent to 'holy' places. Nowadays, there is a general disregard for traditional fishing practices because of the entry of members from ethnic groups without fishing traditions, the attitudes of young fishers who regard taboos as a 'hobby' of older fishers, and primary school dropouts who tend to frown upon traditions.

Source: Tunje (2000)

hexagon patterns. The light weight of the traps, coupled with fairly large mesh sizes made them environmentally friendly (see also Box 3.1). The *uzio* basically consists of a trap enclosure made of sticks tied tightly together. The fences were fixed on the sea floor and aligned perpendicular to the shore extending about ten metres into the water from the beach. *Bunduki* were made of wood and some rubber strips. A metal spike, used to pierce the target, was propelled by a rubber band. It resembled the gun used in sport fishing. The traditional traps and fences have become less popular and were mainly used by a few, older fishers. Spear gun fishers had to be in good physical condition to swim long distances and to hunt moving targets. Nearly all spear gun fishers were younger than 40 years of age.

Table 3.2 Fishing gear by coastal tract (%)*

	CT Ngomeni (N=40)	CT Malindi (N=40)	CT Mida (N=39)	CT Kilifi (N=40)	CT Takaungu (N=40)	Total (N=199)
Beach seines	10.0	–	–	7.5	7.5	5.0
Gill net	90.0	50.0	69.2	77.5	62.5	69.8
Long line	20.0	30.0	35.9	30.0	17.5	26.6
Hand line	57.5	57.5	66.7	32.5	67.5	56.3
Trap	–	25.0	10.3	5.0	2.5	8.5
Fence	–	–	–	2.5	–	0.5
Spear gun	–	–	12.8	–	30.0	8.5
Other	–	–	–	5.0	–	1.0

Source: Fisher survey
* MR (Multiple response)

Modern manufactured gear were industrially produced and had to be purchased. They included all types of nylon nets (gill nets, floating nets, sardine nets, and beach seines) and lines (long lines and hand lines), briefly described below. The floating gill net *(jarife)* had large mesh sizes of more than 4 inches and was used to catch larger pelagic fish, mainly outside the reef. They were normally set overnight and the catch was collected the following day. The gill net (*mpweke*) had a legal mesh size of not less than 1.95 inch to allow small and juvenile fish to pass through. They were used in shallow areas of the lagoon where there were no corals to entangle the nets. The sardine net (*kimia*) had small mesh sizes measuring less than 1.95 inch, and were used mainly to catch small but mature fish such as sardines. The net was cast from a *dau* to surround and enclose fish that were spotted. Beach seines (*juya*) were extremely fine meshed nets prohibited under the Fisheries Act (Kenya 1991). Beach seines not only captured fish of all sizes but also destroyed juvenile fish and larval stages of fish. In addition, they scraped the sea-bottom and physically destroyed corals and reefs. Long lines consisted of a long line floating on the surface to which smaller lines with one or more hooks were attached while hand lines (*mishipi*) consisted of a single line, held by the fisher, with one or two hooks at different depths. Lines were considered non-damaging since they did not damage the reef structure and fish size could be regulated by changing the size of the hooks.

The most popular gear reported by the fishers were the gill nets followed by hand lines, which were in use by more than half the fishers (Table 3.2). Long lines were reported by a quarter of the fishers. The traditional gear were reported less frequently and never by more than 10% of the fishers. Again, there were quite large differences among the coastal tracts depending on the marine environment and the vessels used. As mentioned earlier, fishers generally did not approve of damaging gear such as beach seines, spear guns and poison since they were aware that they will ultimately affect catches negatively (see Box 3.2).

Ethnic tradition in fishing

The coastal population is of mixed origin with the *Mijikenda* being the largest group. These agriculturalists traditionally lived in small villages in the forests on the coastal plateau and coastal range but have moved into the coastal strip in great numbers over the last century and a half. The Mijikenda are subdivided into seven subgroups of which the Giriama is the largest. Since independence in 1963, many immigrants from up-country have also moved to the Coast seeking employment in Mombasa and, lately, in the tourism industry. Traditionally, inhabitants of the coastal strip were the *Swahili* and *Bajun*. The Swahili are regarded as offspring of Arab settlers and indigenous Bantu. They inhabited the 'stone' towns, were involved in trading and enjoyed political dominance (Middleton 2000). The Bajun are regarded as either an independent ethnic group or a sub-group of the Swahili. The Bajun and Swahili are all Muslims; the Mijikenda are mostly Christians (43%) but some are Muslims (29%) or practice African traditional religion (27%).

Bajun

The Bajun were fishers *par excellence*, although the Digo on the south coast have also engaged in fishing for generations. There were also some Swahili although their numbers were small. These groups were coast dwellers of long tradition. The Bajun originated from a group of islands near the Somali border and in recent generations have settled further and further to the south. (The first fisher who settled in Mayungu, one of our research sites, was a Bajun still living there at the time of study and, by now, an old man who was proud to be recognized as the first settler). In the 5 coastal tracts spread over 125 km of coastline, Bajun comprised about 40% and Mijikenda about 60% of the study population.

Mijikenda

Mijikenda have joined the fisheries in large numbers since the 1960s, although this ethnic group does not have a tradition of sea fishing, has little traditional knowledge on how to manage marine resources and does not provide apprenticeships for young fishers (Glaesel 1997a). Nevertheless, they posed considerable competition by now. Many Mijikenda fishers were found in the coastal tracts in Kilifi District and they appeared to be spreading north. In the three northern tracts, Bajun and Mijikenda fishers were present in similar numbers (48% vs. 48%), but in the two southern tracts there were many more Mijikenda fishers (68% vs. 32%). Over the years, the ethnic groups were becoming mixed with Bajun living in Kilifi and Mijikenda fishing as far north as Ras Ngomeni.

Half the Bajun fishers came from fisher families but few of the Mijikenda were second-generation fishers (Table 3.3). For the Bajun, who have relied on fishing or fishing related activities for generations, economic diversification presumably involved a major change in livelihood strategies although there existed some traditional ways of diversification in this group, notably the renting out of houses and farming (Versleijen 2001). The Mijikenda, who came from a farming background were accustomed to economic activities other than farming and, to them, fishing was just one activity among others. Further information on the livelihoods of Bajun and Mijikenda fisher households revealed important differences; differences that were in line with the existing perception of the groups. Half the Mijikenda fishers reported that they had land for farming, but only a few Bajun fishers had land. Mijikenda fishers reported more activity diversification than Bajun, 63% and 18%, respectively, but also more earner diversification, 49% and 29%, respectively. In all, Mijikenda households had higher total incomes but they had to support larger families than the Bajun.

Table 3.3 Selected characteristics of Mijikenda and Bajun fishers and their households

	Mijikenda (N=105)	Bajun (N=28)
Household size	8.3	5.4
2nd-generation fishers	16%	54%
Farmland available	79%	11%
Activity diversification*	63%	18%
Earner diversification*	49%	29%
Household income (Ksh)	2062	1564

Source: Household survey
* For definition of concepts, see Chapter 5

*Wapemba***

These seasonal fishers from the island of Pemba (Tanzania) visited villages such as Mtwapa, Mayungu and Bofa during the high fishing season (September to March) each year. Although it was generally thought that they were in search of better fishing grounds after having exhausted their home grounds (King 2003), their presence was already reported in official documents as early as 1948 (Kenya 1950b). They often operated as legitimate fishers in Kenyan waters because they acquired national identity cards and fishing licenses from the government authorities. They also interacted freely with the local people and married local women so that they could be fully accepted by the host communities.

The favoured fishing gear of the Wapemba was the beach seine. According to members of the local fisher committee, fishing in Mayungu was highly

** This section on the Wapemba is from Tunje (2000).

threatened by the Wapemba because, with their use of unlawful fishing gear within the reefs, they destroyed fish larvae, juvenile fish and corals. Small immature fish, sometimes measuring only three cm and not fit for sale, were discarded. This totalled as much as 100 kg per day. 'Trash-fish' were sometimes buried at the beach, causing more marine environmental pollution (Daily Nation 1998). An area which has been fished over in this way takes approximately 90 days to show the first signs of recovery.

Many local fishers blamed the Wapemba for the decline in fish catches and degradation of fishery resources. Wapemba used *juya* nets although these nets were forbidden and usually managed to avoid authorities. Local fishers found this highly frustrating and claimed that whenever they (local fishers) used illegal gear, they were caught and fined. To protect and conserve their resources, some local fishers staged physical confrontations with these 'foreigners'. This is the reason why the Wapemba were not allowed in some landing sites such as Uyombo and Takaungu, although in other landing sites they were free to operate. In fact, in Mayungu, the local fishers teamed up with them.

In all fairness, the Wapemba should be given some credit for their economic importance to the host communities. Firstly, since they were superior fishers and had all the necessary skills, they facilitated the transfer of fishing technology to the local fishers. Secondly, they created employment opportunities for local youths who did not own fishing equipment. Thirdly, some women could operate as fish traders during this season and thus had a temporary source of income. Lastly, Wapemba fishers could be generous to the local fishers who received free fish for family consumption (sometimes even for sale) when the local fishers failed to go fishing or failed to catch any fish.

Conclusions

This chapter provides a description of the local fisher population. The Fisher Survey covered a random sample of 199 fishers from five coastal tracts that were spread over a distance of about 125 km. It examined a wide range of subjects from fisher organization (crew formation, ownership of equipment) to fishing methods (vessels, gear), fishing grounds, fishing frequency, landing sites, fish landings, catch disposal, and the problems or obstacles fishers experienced. The fisher population comprised boat captains and crewmembers (who accounted for roughly a quarter and two-thirds of the fisher population, respectively) and a small group of independent fishers.

Most fishing efforts were concentrated inshore as is typical of artisanal fisheries. Canoe and dau comprised about 70% of the craft and they were suited mainly for reef and in-reef fishing. Vessels could be owned by one of the fishers but, equally, many captains did not own boats but rented them from a *tajiri*. In some fishing villages, these traders played an important role, elsewhere less so. Modern gear, gill nets and lines, were used most often while traditional gear such as traps and fences were on the decline. The gear were usually owned by the boat captain who was also responsible for the maintenance of the equipment.

Two-thirds of the fishers regularly visited other landing sites which suggested that many of them moved along the coast in search of richer fishing grounds. Fishers sold most of their catch and took some fish home. Two-thirds of the fishers sold the fish in fresh condition but, in remote landing sites, the fish were commonly dried or fried. Lack of equipment was most often mentioned as a problem, followed by lack of funds, and transport and marketing bottlenecks, although there were considerable local differences in the kinds of obstacles. Some of the site-specific problems were the competition by trawlers and the presence of nearby Marine Protected Areas.

There were large differences among the coastal tracts in terms of vessels and gear as well as large differences in fishing practices among individual fishers at the same site.

Table 3.4 Vessels in use by artisanal fishers, Kenyan coast

	Length	Clearance	Construction	Propulsion	Centre pole (for sail)	Steering	Rudder fixture	Crew	Price [7] (Ksh - 2004)
Mtumbwi	3 m	Low	Dug-Out	Paddles/Poles; few sails	No	Paddles Poles	n.a.	1-2	7,000
Hori	3 m	Low	Plank	Sail	Removable	Rudder with shoulder rope	Removable	2-3	13,000
Ngalawa [1]	5-6 m	Higher than Dau	Dug-Out	Sail	Permanent	Rudder with stick	Removable	4	40,000
Dau [2]	5 m	Low [5]	Plank	Sail	Removable	Rudder with stick	Permanent	2-3	25,000
Mashua	10 m	High + Talbisi [6]	Plank	Sail	Permanent	Rudder with stick	Removable	4-6	250,000
Jahazi [3 4]	20 m	High + Talbisi	Plank	Sail	Permanent	Rudder with stick/wheel	Permanent	10+	++

1. *Ngalawa* are not made locally, but come from Pemba.
2. *Hori* and *dau* are sometimes difficult to distinguish but *dau* are wider inside, have a flat bottom, are longer, and have a curved silhouette.
3. A *jahazi* is technically used to transport goods, not for fishing. Nowadays the terms are used loosely and a large *mashua* may be called *jahazi*.
4. Sometime *jahazi* carry a smaller *mashua* inside as a lifeboat or landing craft.
5. *Dau* clearance is low because it is used for collecting traps.
6. *Talbisi* are mattings used to increase clearance above the water.
7. Price depends on the size; whether hardwood or softwood; the boat builder or *fundi* (the ones from Lamu are the best); and the sail.

Table 3.5 Gear in use by artisanal fishers, Kenyan coast

	Swahili	Deployment	Day-Night	Min. crew	Boat type used	Manufacturing price
Trap	Malema	Kusi: First Reef; Kaskazi; Second Reef; Opening facing current; Stones on top; Positioned by skin divers; Age no constraint	D	2	Dau	owner-made; baited with seaweed and small crabs that are crushed together
Fence	Uzio		D/N (when moon is full)	1	=	owner-made
Stick	Mkonjo		D/N	1	Foot	owner-made
Spear gun	Bunduki		D/N	Foot (1) Boat (2)	Hori Mtumbwi	owner-made
Poison	Mchupa	Once the tide is out: Mayungu Ras ya Ngome Takaungu	D	1	=	collecting; Mchupa Tree; scratch the bark
Poisonous stick	Mchupa		D	1	=	collecting; Branch of Mchupa Tree

Table 3.5 Fishing gear, *continued*

	Swahili	Size	Deployment	Day-Night	Min. crew	Boat type used	Manufacturing price	Cost of extras
Hand line [1]	Mshipi	10-15 m deep	1 or 2 hooks at different depths (deep & medium); Can be used stationary (anchor) or trawling (in that case only 1 hook)	D/N	1	Mtumbwi Hori Ngalawa Mashua (small)	100 yards; 1.20 mm Ksh 180	2 hooks No.4 (Ksh 30 each) 2 weight (Ksh 10 each)
Long line	Long line	50-100 m long 10-15 m deep 1 subline/ 2 m 1 hook per sub-line;		D/N	3	Dingi Mashua (+ engine)	100 m Ksh 3,000 (50 m for main line; 50 m for sublines)	25 hooks No.2 (Ksh 20 each) 2 Anchors (Ksh 1,500 each) Buoy/Rope (Ksh 1,000)

1. Also called Hook and Line

Table 3.5 Fishing gear, *continued*

	Swahili	Size	Mesh size	Deployment	Day-Night	Min. crew	Boat type used	Manufacturing price	Cost of extras
Gill net: By hand	Mpweke	2 m deep 8 m long no trawling	2.0 inch	Mayungu; Uyombo Used to encircle near reef, with beaters and attached to foot.	D/N	3	Mtumbi Hori	50 yards Ksh 750	floaters/rope/ threads/weights Ksh 2,500
Gill net: Surface floating[2]	Jarife	4-5 m deep 8 m long Up to 5 pieces can be connected	4-6 inch	Done at nights without moon but with a light (pressure lamp). Boat and net float with current. Harvest morning	N	4	Mashua (big)	100 yards Ksh 20,000	floaters/ropes/ weights Ksh 4,000
Gill net: Bottom fixed[3]	Jarife	4-5 m deep 8 m long Up to 5 pieces can be connected	4-6 inch	Done at nights with moon, fixed place, inspect morning, evening	N	4	Mashua (big)	100 yards Ksh 20,000	floaters/ropes/ weights Ksh 4,000
Beach seine	Juya	2 m deep or less Length; <8 m trawling	1 inch		D/N	2	=	50 yards Ksh 1,000	floaters/ropes Ksh 1,000

2. Also called floating net
3. Also called fixed net

Table 3.5 Fishing gear, *continued*

	Swahili	Size	Mesh size	Deployment	Day-Night	Min. crew	Boat type used	Manufacturing price	Cost of extras
Ring net[4] Recent method	No Swahili	200 m long 10-15 m deep Sometimes with divers, to close the net below	1 inch	Outside 3 nautical miles. Day: Locate shoal, 2 motorboats encircle shoal with nets. Nights: Pressure Lamp, wait for shoal - continue	D/N (mostly night)	8	2 motor boats (dingi or mashua)	400 yards Ksh 200,000 - 300,000	(all inclusive)
Trawling net: Hand		10-15 m deep long: 100 m trawling	1 inch; thick thread	Used with 1 boat to encircle outside the reef; Need many people	D/N	20	Mashua (big)	200 yards Ksh 40,000-50,000	(all inclusive)
Trawling net: Commercial				Trawlers					
Scoup net	Senga			Used to lift a catch by line fishers or squids (as bait) attracted by pressure lamp					

4. Ringnets may trap large amounts of fish (3-5 tons is possible) but they have been banned as from April 30, 2006, because of their small mesh size.

Figure 4.1 Quantities of Marine Landings, 1956-2005
(1956-1964 Combined catches; 1965-2005 Breakdown by finfish, crustaceans and other marine organisms)

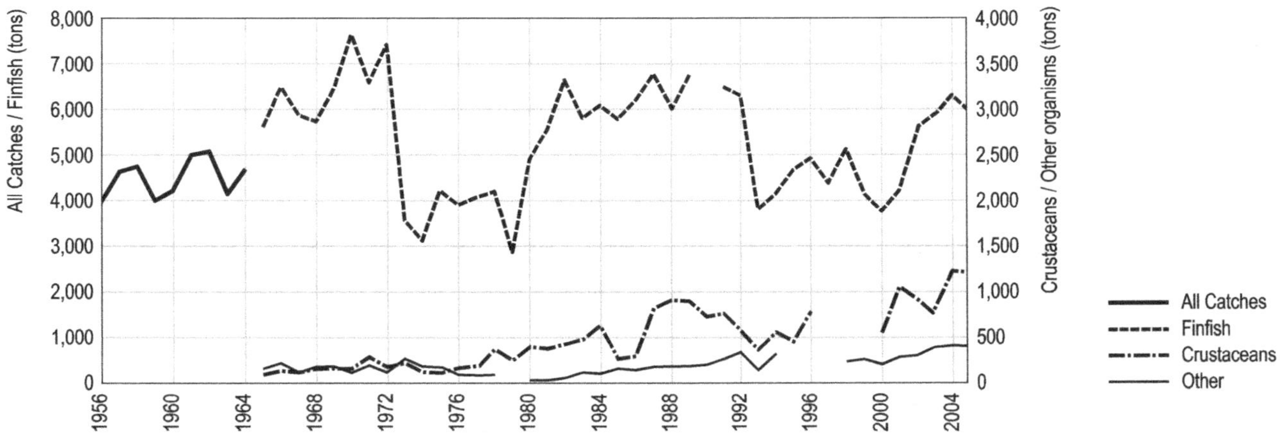

Source: 1956-1964: Report of Kenya Fisheries (Kenya 1957-1966); 1965-2005: Statistical Abstract (Kenya 1970-2007)
(Interrupted lines for certain years indicate that values for these years are missing or considered to be unreliable)

4

Fish landings

Fish species composition and catch size*

In 2005, the quantity of fish landed in Kenya was 146,642 metric tons, an almost five-fold increase over the 29,808 tons in 1972. This increase was due to the growth of freshwater fisheries, which accounted for 95% of the fish caught in the country (Lake Victoria fisheries alone contributing 90%). The value of all fish landed to the fishers in 2005 was Ksh 7,974 million or US$ 110.2 million (Kenya 1975, 2007a).

Marine catches of all categories (including finfish, crustaceans and other marine organisms) in 2005 were 7,616 tons compared with 7,722 tons in 1972, both being good years (Kenya 1975, 2007a). The value of marine catches to the fishers in 2005 was Ksh 587.2 million (US$ 8.1 million). However, large variations occurred among years, particularly when sub-categories were examined. Figure 4.1 shows that the landings of finfish fluctuated considerably with periods of high catches (1965-72; 1982-92) alternating with low catches (1973-79; 1993-2002). In contrast, the production of crustaceans (lobster, prawn, crab) has shown an overall increase over these years, although there were short dips in 1985-86 and 1992-95.

---

* The findings in this section are from the Catch Survey at four landing sites unless indicated otherwise.

The highest catch of finfish was recorded in 1970 with 7,616 tons; the lowest in 1979 with less than 3,000 tons. During the last decade, catches of finfish have been generally on the low side. The average catch between 1993 and 2002 was 4,357 tons although the most recent figures showed a recovery. The catch of crustaceans has surpassed 1,000 tons (Figure 4.1).

In 1972, finfish comprised 96% of the marine catch but in 2005 its contribution was lowered to 79%. Crustaceans and other marine organisms (oyster, bêche-de-mer, octopus, squid) contributed 16% and 5%, respectively. In money terms, however, the breakdown was rather different with finfish accounting for 56% of the product value, crustaceans (39%) and other marine organisms (5%). It is evident that crustaceans are becoming more important in the fishery and may soon overtake the proceeds realized by marine fishers.

There were 63 fish species recorded during the 18 months Catch Survey but 104 species were recorded during the 6 months support study on fish landings and reproductive biology. The difference occurred not because more species were caught but because the latter study made a conscious effort to identify species more finely (for example, seven different types of goatfish vs. goatfish in general). The combined results list 113 species (Appendix 1, p. 118).

Table 4.1 Fish species composition by landing site (%; number of times listed species was present in catch records; 5% or more).

	LS Ngomeni (N=1227)*	LS Mayungu (N=1516)	LS Uyombo (N=1751)	LS Takaungu (N=942)	Total (N=5436)
Rabbitfish	3.4	59.0	75.8	17.4	44.6
Emperor	2.1	19.3	64.7	21.8	30.5
Parrotfish	0.4	1.6	32.6	3.2	11.6
Snapper	2.9	1.6	23.4	16.8	11.5
Goatfish	-	0.1	33.1	0.4	10.8
Wrasse	-	0.2	28.7	0.2	9.3
Mullet	35.5	0.1	0.7	3.9	8.9
Rock cod	2.8	1.6	15.2	4.4	6.7
Kingfish	12.8	3.1	0.7	10.8	5.8
Ribbonfish	0.1	0.1	0.1	30.2	5.3

Source: Catch Survey.
* N = Number of weighted catch records
Further data on catch composition are listed in Appendix 2, p. 122.

The demersal or bottom-dwelling fish dominated and accounted for about 40% of the total marine catches (Mohammed 2002). More than fifty species of demersal fish were identified, the most important of which were rabbitfish (*Siganidae*), scavenger (*Lethrinidae*), snapper (*Lutjanidae*), grunters (*Pomasyidae*), rock cod (*Serranidae*) and parrotfish (*Scaridae*). The pelagic or surface-dwelling fish contributed approximately 35%, with at least thirty different species. Included in this category were cavilla jack (*Carangidae*), barracuda (*Sphyraenidae*), mullet (*Mugillidae*), bonito (*Scombridae*), sailfish (*Istiophorus*) and kingfish (*Menticirrhus*). Shark, ray and sardine accounted for 9% of the total catch in 1999.

Many of the species occurred only occasionally; only 10 species were present and reported in more than 5% of all landings. From the breakdown by landing sites, it was evident that the differences in catch composition among sites were large, not only between Ngomeni and the others but also among the landing sites situated near the reef (Table 4.1).

Mullet was the fish most commonly reported from the Ngomeni LS, which had quite a different ecology from the other landing sites. There was no fringing reef but open sea with waves rolling in from the Indian Ocean, long sandy beaches and mudflats with mangroves and salt farms. Kingfish, shark and lobster were also frequent in Ngomeni. In the Mayungu LS and Uyombo LS, rabbitfish was most prevalent, being present in half to three-quarters of the catches (see Box 4.1). In Mayungu only one other species was frequent, the emperor. In Uyombo rabbitfish and emperor were the most common but parrotfish, snapper, goatfish, wrasse, rock cod and sweetlips were also common. This landing site was next to a Marine Protected Area, where a large creek and the sea met. In Takaungu LS, rabbitfish and emperor were, again, prevalent but ribbonfish, a fish that swarms and breeds in this area, were reported most often. In Takaungu, the habitat consisted of the offshore coral reef and a narrow, deep creek but there was no Marine Protected Area nearby. Also present here were snapper and kingfish (Table 4.1). The number of species was lowest at Ngomeni (28 species) followed by Takaungu (31) and Mayungu (35) and highest at Uyombo (47), next to the Watamu Marine Park (Table 4.2).

Box 4.1 Rabbitfish under threat

The rabbitfish (*Siganus sutor*) forms the largest part, by far, of the artisanal landings along Malindi-Kilifi inshore waters (Ntiba 1986). Kaunda-Arara (1997) noted that this species is subjected to intensive fishing. Studies of the reproductive biology of marine fish in the coastal waters of Kenya were already available for monocle bream, *Scolopsis bimaculatus* (Nzioka 1981), rabbitfish (Ntiba 1986) and snapper, *Lutjanus fulviflamma* (Kaunda-Arara & Ntiba 1997). One of our support studies further examined the reproductive biology of the rabbitfish.

The overall sex ratio of the population studied was not significantly different from the expected 1:1. The fecundity of *S. sutor* ranged from 170,000 eggs in a female measuring 17.5 cm (TL) and weighing 80 g to 781,000 eggs in a fish of 24.0 cm (TL) and weighing 225 g. The mean fecundity was estimated at 506,000 ± 30,327 eggs. The gonads attained a peak weight at stage IV and then a gradual decrease in weight through stage V, to stage IIb. The mean gonadosomatic-index (GSI) markedly dropped between the months of January and February. The highest mean GSI for males was recorded in November while for females it was in December.

Size at first maturity is another important indicator of the status of the population. The minimum length at which 50% of the females contained ripe eggs was 17.5 cm TL (12.8 cm standard length from snout to beginning of caudal fin (SL) with a corresponding weight of 85 g. The smallest mature female measured 16.5 cm TL. Beyond the length class of 17.0-19.0 cm all females had passed through stage II of maturation. Spent females were found in the samples from the range of 26.0 cm and above. About 50% of mature males had a minimum length of 17.0 cm TL (12.2 cm SL) with a corresponding minimum weight of 90 g. The length 17.0 cm also represented the smallest mature male.

The size at first maturity represented a reduction from 18.0 cm TL for both females and male, reported by Ntiba (1986), and suggests that fish are maturing at an earlier age. This confirms that the rabbitfish population is under pressure and that its reproductive biology is most likely affected by excessive fishing.

Source: Mohammed 2002

The average weight of the catches showed pronounced differences among landing sites (Table 4.2). Uyombo, with the largest species variety, reported the smallest catches with 50% below 4.0 kg. Catches in Mayungu were larger than in Uyombo (average 9.7 vs. 4.7 kg), but they still were substantially lower than in Takaungu and Ngomeni (average of 18.8 and 25.6 kg, respectively). The explanation was that the fisheries in Ngomeni and Takaungu were of a different nature, with larger vessels for the open sea, larger crews and, consequently, larger fish caught.

Catches translated into financial earnings which depended on the amount and type of fish landed (for most fish species there was an agreed price), the number of crew who divided the catch and other shares for vessel, gear and captaincy. Most of the earnings per crewmember per trip were modest; the overall average was Ksh 372 and 50% was below Ksh 250 (Table 4.2). The lowest earnings per trip were reported from Uyombo, followed by Mayungu. These two areas had the lowest weights landed and were situated in Marine Protected Areas with regular patrols by KWS wardens. Earnings per trip in Ngomeni were considerably higher than in Mayungu and Uyombo and corresponded with larger catches while the highest earnings were reported from Takaungu.

Catches were highest during the *kaskazi* season as already reported by Nzioka (1984) and McClanahan (1988). Figure 4.2 presents a seasonal overview of species diversity, catch weights and incomes for the whole study area. Species diversity varied from a low of 34 in May to a high of 48 in November. Weight caught per trip showed a pattern of high catches from October to April and low catches during July, August and September. From these data, it was not possible to determine whether the differences were the results of changes in the presence of species due to weather conditions or due to changes in fishing practices between seasons. Not surprisingly, earnings showed the same seasonal pattern with low incomes from July to September.

Table 4.2 Catch characteristics by landing site.

	LS Ngomeni (N=1227)*	LS Mayungu (N=1516)	LS Uyombo (N=1751)	LS Takaungu (N=942)	Total (N=5436)
No. species 1	28	35	47	31	63
Catch weight 2	25.6 (17.0)	9.7 (9.9)	4.7 (5.1)	18.8 (17.8)	13.3 (15.0)
Earnings 3	580 (381)	298 (299)	158 (125)	692 (592)	372 (394)

Source: Catch Survey: species that were present in 0.5% of the catch records or more.
* N=Number of weighted catch records
1) Total number of species recorded
2) Average catch weight (kg)/Trip (s.d.)
3) Average earnings (Ksh)/Crewmember/Trip (s.d.)

Figure 4.2 Catch Characteristics by Season (Source: Catch Survey)

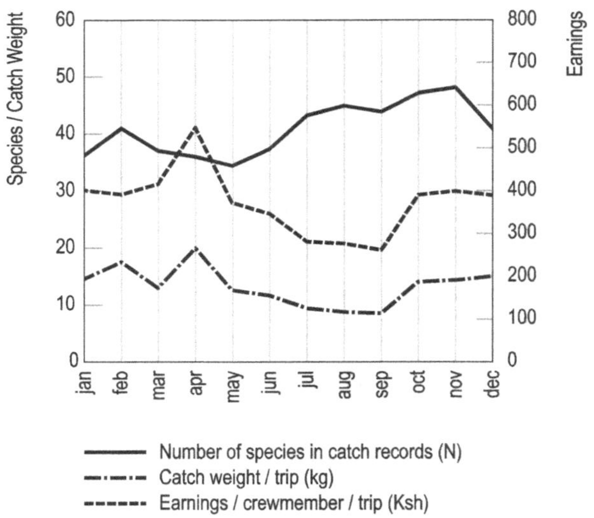

Breakdown by landing sites revealed a more complex picture (Figure 4.3). In Ngomeni, mullet, the leading fish caught in this area, had a relatively low presence from April to June. During that period, however, kingfish and shark were reported more frequently. In Mayungu, rabbitfish and emperor showed distinct seasonal patterns with low catches in July to September, and there was no compensatory trend noted. In Uyombo, rabbitfish, the most common catch, was least frequent from January to March and the second most common, emperor, showed essentially the same pattern. Goatfish and parrotfish, also showed seasonal variation with low presence in catch records from July to September. Finally, in Takaungu, ribbonfish dominated the catch from June to September. Emperor showed essentially the opposite (compensatory) pattern while rabbitfish and snapper were low from August to December. Again, from these trends, it was not clear whether the respective species were less or more abundant during these periods or whether the trends reflected changes in fishing behaviour of the fishers. (Further specification of catch data are given in Wangila, Hoorweg & Degen 2007a).

In Uyombo, the catch weights landed were very low and varied between an average of 3.5 and 6.2 kg throughout the year. Catches in Mayungu were also low but there was a distinct seasonal pattern with best catches from October to February. In Ngomeni, catches were low during two periods, namely December to January and July to September; during the rest of the year there were considerable fluctuations. The Takaungu catch patterns fluctuated most. Catches were low in the period July to November but rose steeply to an average of 55 kg in April. Outliers did not cause these high figures, as many fishers reported high catches during that month, but this was partly the consequence of vessels used at this site, notably motorboats.

Earnings per trip corresponded with the catch weight trends and they were lowest in Uyombo throughout the year. Earnings in Mayungu were also low during much of the year but there was an increase from October to January. Despite a species-rich catch, the total weight per trip and income were nevertheless low in these areas. On the other hand, those fishing in unprotected waters were better off (see Wangila *et al.* 2007a). Earnings in Ngomeni were higher with monthly fluctuations, being lowest from July to September. The pattern in Takaungu was skewed with high earnings from February to May but low from June to November, when they were at the same level as in Uyombo and Ngomeni.

Fisher incomes

Weekly incomes and income composition of the fishers were examined in four landing sites as part of the Household Survey (see Chapter 2). The sites offered different combinations of geographical factors that might potentially affect income levels, namely (i) access to urban centres with employment opportunities and (ii) nearby presence of a Marine Protected Area (Table 4.3). Fishers in remote locations (Ngomeni and Uyombo) had the lowest total income. At first glance, there was no clear relation between incomes and presence or absence of protected areas since the two locations that were situated at the Malindi/Watamu Marine Reserve had not only the highest (Mayungu) but also the lowest incomes (Uyombo). Further breakdown of income is presented in terms of fishing income and incomes from farming and employment.

Figure 4.3 Catch Composition by Landing Site by Season for the Four Main Fish Species at the Site
(N=number of times that species was reported in catch records) *Source:* Catch Survey

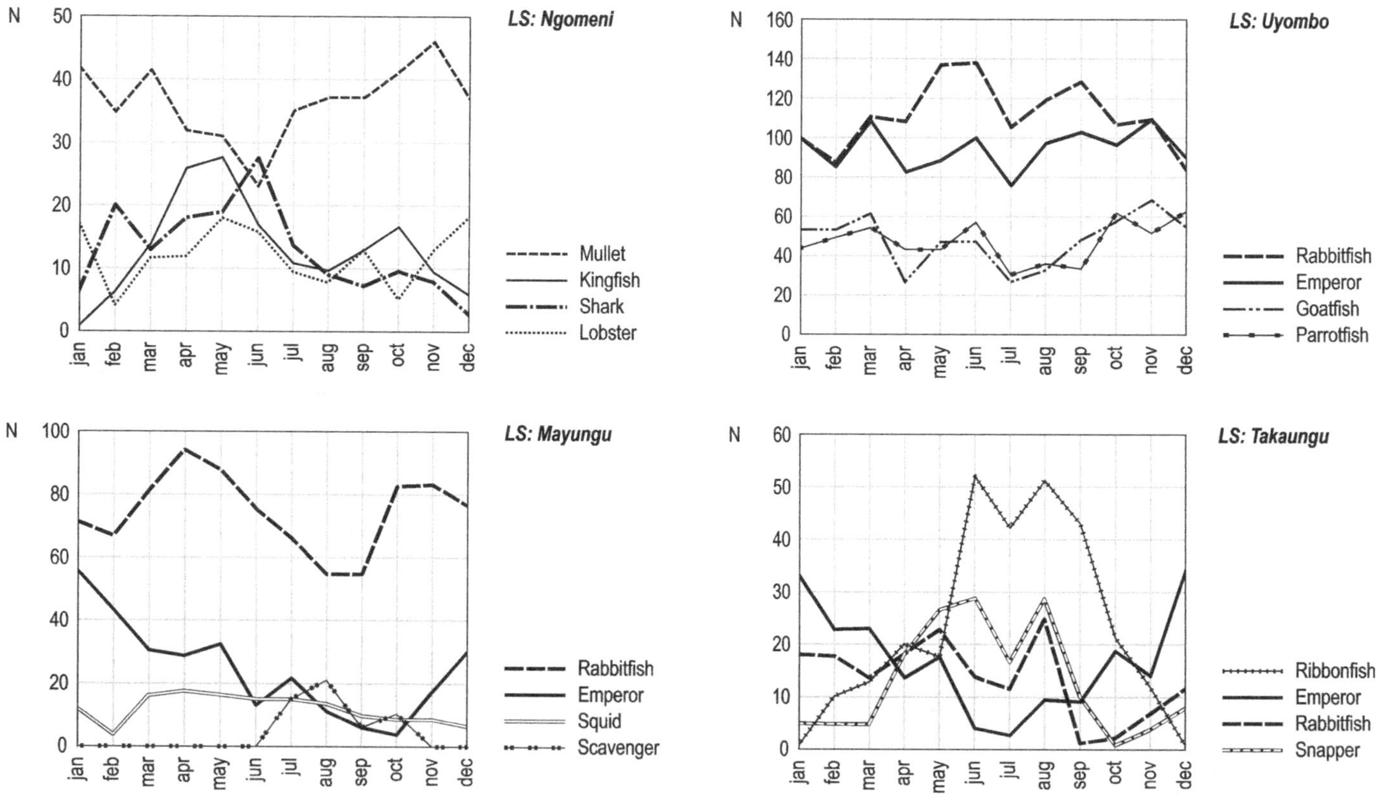

The lowest fishing income was realized by fishers in Uyombo and this agreed with the fact that the earnings per trip (Table 4.2) were the lowest of the four sites. The highest fishing income was reported by fishers in Mayungu and this was so despite the fact that the earnings per trip at this site were low, although not as low as in Uyombo. This discrepancy between earnings per trip and income per week in Mayungu could be related to several factors. Firstly, fishers in Mayungu had more fishing trips, although the difference with the other locations was not large enough to explain all of the difference. Secondly, fishers often fished elsewhere and/or frequented other landing sites; it is likely that in this case, the fishers from Mayungu brought large catches to Malindi where they were more likely to find buyers and brought only the smaller catches to Mayungu LS itself. A third factor could be the seasonal presence of Wapemba fishers in Mayungu from which the local fishers also benefited (see Chapter 3).

Table 4.3 Income composition of fishers by landing site

	Nearness urban centre	Nearness to MPA*	Income fishing	Income farming	Income employ-ment	Income total (s.d.)
LS Ngomeni	–	–	1504	52	0	1556 (905)
LS Mayungu	+	+	1577	180	71	1828 (1291)
LS Uyombo	–	+	1101	325	65	1491 (1009)
LS Takaungu	+	–	1357	215	174	1746 (999)

Source: Household Survey
* Marine Protected Area

The fishers in Uyombo appeared to be in an unfavourable position. They reported fewer fishing trips than fishers in Mayungu and it was more difficult for them to land their catch in Malindi because of the larger distance. In addition, they were hostile to the Wapemba fishers and were not in a position to benefit from the activities of these versatile fishers. The fishing incomes of fishers in Ngomeni and Takaungu, neither located near a Marine Protected Area, were similar despite the differences in geographical conditions of the sites and differences in types of fish species caught.

Incomes from fishing were strengthened to a greater or lesser extent with incomes from agriculture and employment. The agricultural potential in the Ngomeni area was less than in the other sites and the fishers here earned very little from agriculture. Income from agriculture was higher in the other sites. Earnings from agriculture were highest in Uyombo and it is very likely that the low income from fishing forced people to put in more effort in farming.

Incomes from employment were generally small. Fishers had limited time, not only restricting the possibilities to do other work but certainly restricting the possibilities to commute to a place of work elsewhere, which is often necessary to find employment. In Ngomeni, fishers did not have employment income at all; there were hardly any employment opportunities at the site and none were available in the near vicinity as well. Fishers in Mayungu and Uyombo lived in less isolated conditions although job opportunities were still few and incomes from employment were relatively small. In Takaungu, there were some employment opportunities and there was regular public transport to Kilifi and Mombasa. It was understandable, therefore, that fishers here had the highest employment incomes.

The income figures were restricted to the incomes of the fishers, but to obtain a true picture of household resources, the economic activities and incomes from other household members also were examined. Information on the contributions of wives and children is presented in Chapter 5, which compares the incomes of fisher households with non-fisher households with focus on different types of fishers and differences in kind of income diversification.

Fish handling and marketing*

Once the catch was landed, the fish were sold to various traders. Almost all marine fish marketing in Malindi and Kilifi Districts was on a local scale. This type of marketing involved buying and selling at landing sites, open markets or fish shops. The sellers included both fishers and traders in their individual capacities. Most of the fish was sold fresh and consumed locally, although some dealers took them to Mombasa, Nairobi and elsewhere. The fish were nearly always sold to traders and intermediaries, hardly ever directly to consumers. In about a third of the cases they were sold to a *tajiri* who had the right of first re-

* The findings in this section are from the Trader Survey at five coastal tracts unless indicated otherwise.

fusal for the catch of certain fishers; clients who rented his boat or who had been assisted in some way in the past. This occurred more often in the coastal tracts of Ngomeni, Kilifi and Takaungu. At least half the traders were women.

Traders usually had a specialty such as small finfish (29%), medium finfish (35%), large finfish (18%) or shellfish (14%). They differed significantly in the average amount of fish they purchased daily; roughly a third bought less than 10 kg (small-scale buyers), a third between 10 and 50 kg (medium) and a third more than 50 kg (large-scale). Half these traders had more than five years experience in the trade. Most traders frequented only one landing site while almost 40% of the traders frequented two, three or even more sites. Only a third of the traders offered incentives such as financial assistance (15%) or foodstuffs (9%) to the fishers during times of hardship. Almost half the traders, fried the fish before selling them. Only a third of the traders had access to a cooler or freezer. The destination of the fish included the nearest village (29% of the traders), the nearest village on the tarmac road (27%) and urban destinations such as Malindi or Mombasa. Transport was either on foot (43%), bicycle (23%) or matatu/bus (33%).

Table 4.4 Finfish buying price by sales category and coastal tract (Ksh/kg; ave.).

	N	CT Ngomeni	CT Malindi	CT Mida	CT Kilifi	CT Takaungu
Small fish	102	66.0	69.2	62.5	58.9	44.4
Medium (B)	103	54.6	65.3	67.8	75.7	55.0
Medium (A)	120	57.3	71.3	67.7	77.4	66.6
Large (B)	62	90.5	68.3	72.3	80.0	65.0
Large (A)	69	94.6	70.0	72.3	85.0	93.3

Source: Trader Survey

For pricing purposes, finfish were divided into five categories, namely small (<20 cm), medium (20-50 cm) and large fish (>50 cm) with the medium and large fish being further divided into Grade A and Grade B, depending on the degree of freshness ("some fish are more dead than others"). Fixed prices for the five categories were generally accepted with possible differences among landing sites (depending on accessibility and transport costs). Fishers and traders rarely argued about the price for a certain category of fish, but arguments did occur about the size classification of the catch and between grade A and B. The buying prices and selling prices reported by the traders for the five categories of fish were recorded. On average, there was a price difference of Ksh 5/kg between the different categories of fish, starting from about Ksh 60 for small fish to Ksh 80 for large (A) fish.

Essentially prices were decided by two factors, fish size and location of the landing site, but breakdown by site revealed some further variation (Table 4.4). Notable were the higher prices that were paid for large fish in Ngomeni and Takaungu compared with the other sites. It is possible that this was caused by large fish such as tuna and sailfish at these sites that were bought up directly for gourmet markets. Such fish were mostly absent in Mida and Kilifi and, if caught, not easily disposed. In Malindi, the price of large fish was depressed because the market was affected by the many sport vessels that landed their catch here. At the other end, the prices for small fish in Takaungu were considerably below the average of the other sites and this was most likely related to the large seasonal catches of low-priced ribbon fish.

Table 4.5 lists the average price for catch categories and for landing sites together with the selling margins that the traders mentioned (the latter is expressed as a percentage of the buying price). It is clear that there were differences in buying prices among coastal tracts. The tracts that were difficult to access had selling margins of about 40%. The selling margins for the Kilifi coastal tract, near Kilifi town, where transport costs were less, was indeed lower (26%). Surprisingly, this was not the case at the sites near Malindi town and it appears

Table 4.5 Finfish buying price (Ksh/kg) and selling margin (%) by coastal tract and by sales category

*Coastal tract**	buying price	selling margin	*Finfish category***	buying price	selling margin
Ngomeni	72.6	+40	Small Fish	62.2	+32
Malindi	68.8	+41	Medium (B)	65.9	+37
Mida	68.5	+40	Medium (A)	70.0	+35
Kilifi	75.4	+26	Large (B)	75.2	+40
Takaungu	64.9	+42	Large (A)	80.0	+37

Source: Trader Survey

* Average for coastal tracts calculated independent of the number of traders interviewed at each site i.e. by adding the averages for different finfish categories divided by five.

** Averages for finfish categories calculated as average of all observations that the market was functioning imperfectly with producer prices being kept low and selling margins high. The selling margins for the different fish categories increased from 32% to 40% depending on the size of fish.[13]

The trader population consisted of about half male and half female traders, with profound gender differences that are discussed in detail in Wangila, Hoorweg & Degen (2007b). A summary of the most important characteristics is given below (Table 4.6). Male traders were from mixed ethnic backgrounds. Many frequented more than one landing site and bought from six fishers or more. They bought all types of fish, from small to large sized fish as well as crustaceans and octopus, and typically bought more than 20 kg, half of them more than 50 kg. Occasionally they offered incentives to the fishers. Generally, they did not process the fish but had access to a cooler or freezer. They sold in various places including nearby villages and urban centres and many transported the produce by mechanized vehicle. Few of the male traders were small-scale buyers during the high season, less than a third medium and 60% large-scale buyers.

Table 4.6 Selected characteristics of male and female traders

	Male traders	Female traders
Number of landing sites frequented	2 or more	1 only
Number of fisher suppliers	6 or more fishers	1-5 fishers
Type of fish specialisation	all types	small + medium sized fish
Daily fish purchases	50 kg or more	20 kg or less
Additional incentives	yes	no
Fish processing	no	frying
Fish storage	yes	no
Nearest sales location	rural & urban	rural
Mode of transport	mechanised	foot
Average trade income (Ksh/week; s.d.)	1693 (1001)	795 (674)

Source: Trader Survey

Women operated almost exclusively as small-scale traders who bought and processed mostly small fish and sold locally. They were typically of Mijikenda origin and had not been in the trade for very long. They frequented only one landing site and dealt with a few fishers that supplied them. They usually bought less than 20 kg/day during the high season, half even less than ten kg/day. They did not offer additional incentives to fishers except occasional foodstuffs. Nearly all women traders fried the fish they purchased and did not use any other means of preservation or storage. The next selling location was either the nearest village or the nearest village on the tarmac road where the fish was taken by foot.

For male traders, the fish trade was their major activity and 50% reported that they had no additional economic activities. For a quarter of the female traders, the fish trade was their only activity while more than half reported they also did farming. In line with the description so far, the average income of female traders from fish trading was about Ksh 800/week with 70% making less than Ksh 1,000/week. For male traders, the average income from fish trading was about Ksh 1,700, which is higher than the average income of about Ksh 1,400 for fishers from fishing (see Table 5.4, p. 74). However, almost half the male traders earned more than Ksh 2,000/week and 20% earned more than Ksh 2,500/week.

Marketing constraints*

Marine fisheries as an economic sector and fishers as individuals faced various marketing constraints. The constraint that was most often mentioned was the lack of storage facilities for the highly perishable commodity (Kenya 1997); which calls to mind the same observation made by the coastal fish warden, Allfree, in the 1950's (see Chapter 1). This often led to the disposal of fish at throw-away prices in order to reduce losses (alternatively, the fish were given to relatives or business contacts of the traders). This was further compounded by the general poor state of the roads leading to the landing sites. As a result, many traders had to incur high transport costs to reach landing sites and markets.

The marine fisheries sector also suffered from lack of credit facilities (Kenya 1997) which negatively affected the development of the sector. Availability of credit ensures that traders can invest in processing, storage and transport facilities. FAO (1996) already advocated the need for value added production in artisanal fisheries but also the need to develop non-traditional fish products. In Ma-

* This section is a selection of the constraints identified by Wamukota (2002).

lindi and Kilifi districts, the prevalent modes of fish processing or preserving included frying, smoking, sun-drying, and salting, all of which were traditional and added little value to the catch.

The failure of many of the co-operative societies also affected marine fish marketing negatively. Some local co-operatives collapsed because of improper management by poorly educated officials who had low managerial skills and lack of foresight, not to mention greed and corruption (Mwakilenge 1996). The government identified the need for education and training as essential to the development of co-operatives and organized a number of courses to train secretaries, managers, chairmen and the general membership but this effort did not yield encouraging results as the cooperatives did not recover. Illiteracy among the fishers and traders has traditionally been a problem of the sector (Kenya 1957; Obura *et al.* 2002). Lack of understanding and information caused the fisher loan scheme that was started in the late sixties to fail in Malindi because it was mistakenly regarded as a grant by the recipients (Mwakilenge 1996).

Conclusions

This chapter provides further information on the nature of the fisheries along the Malindi Coast with details of fish catches and fish marketing. More than 100 fish species were identified at four selected landing sites. Species richness was highest in the landing sites situated near Marine Protected Areas. The amounts of fish per fishing trip landed at these two sites were, however, considerably lower than at the sites that were not near protected areas. Returns per crewmember per trip reflected the same differences, being highest in the non-protected areas and far lower in the protected sites. This confirmed earlier observations that Marine Protected Areas can result in greater fish density and spillover to adjacent waters (Kaunda-Arara & Rose 2004a, 2004b) but that more fishers were concentrated in smaller areas, resulting in lower catches per fisher (McClanahan & Mangi 2000).

Apart from the differences among landing sites, there were also considerable seasonal differences in catch. Diversity ranged from 34 species in May to 48 species in November. The catches and incomes were generally low in the period

July to September but higher in the period October to April. The average earnings per crewmember per trip in the lowest quarter (July to September) was only 60% of that in the highest quarter (February to April). Earnings showed large differences between seasons, among sites and among individuals. Cinner & McClanahan (2006) also reported large socio-economic differences among fishing communities along the Kenyan Coast.

Total income of fishers, which included earnings from fishing, farming and employment, was lowest in remote locations. In addition, fishers next to the Marine Protected Areas showed large differences in incomes from fishing. In Uyombo, the landing site where earnings per trip were lowest, the fishers also reported the lowest weekly incomes from fishing. But, in Mayungu, where earnings per trip were also low, fishers, nevertheless, had much higher weekly incomes from fishing than in Uyombo. Possible reasons were the higher number of fishing trips in Mayungu, transfer of large catches to other landing sites and benefits from cooperation with Wapemba fishers. Income from agriculture and employment also varied among landing sites, depending on agricultural potential of the area and employment opportunities, among others.

Buying prices depended mostly on two factors, landing site and fish size. Five fish categories were distinguished: small, medium (B quality), medium (A), large (B) and large (A) with price increases of about Ksh 5/kg between two categories. The selling margins for the different fish categories varied between 30 and 40%. Half the traders were women, half were men. Most female traders bought small fish and fried them before selling to mostly local customers. The larger fish were the domain of the male traders who did not process the fish but sold them either the same day or stored them in a cooler or freezer.

5

Fisher livelihoods*

Kenya belongs to the bottom 20% of countries in the world in terms of economic and human development. In 1998, development indicators showed that of 174 countries the country ranked 155 in gross domestic product per capita, and 138 in terms of human development (UNDP 2000). In 1997, about half the population of Kenya (52%) was termed poor by national standards and the number was on the increase (Kenya 2001).[14] Per capita income of Kenya is higher than that of Uganda and Tanzania, its immediate neighbours, but Kenya has a more unequal distribution that these two countries (Ellis & Freeman 2004). Among the poorest regions of the country were North-Eastern, Nyanza and Coast Provinces, with 64% of the rural population in the latter termed as poor.[15] Mombasa, with about a third of the population of Coast Province, however, had the lowest percentage of urban poor (38%) of the five major cities in the country (Kenya 2001). This disparity between the rural and urban areas of Coast Province was noted before (Ikiara 2000) and the reasons for the poverty in the rural areas include the climate and poor agricultural conditions, lack of employment opportunities and low levels of education (Hoorweg *et al.* 2000). The backwardness of the rural areas is aggravated by the 'ribbon' type of set-

* The findings in this chapter are from the Household Survey at four landing sites unless indicated otherwise.

tlement, meaning that most people live relatively far from Mombasa, the main centre of economic and employment opportunities. Employment creation throughout the province has been identified as an urgent development priority (Masai 2000).

Marine fisheries are one of the few economic activities present everywhere along the coast but the general opinion is that catches are on the decline (Kaunda-Arare *et al.* 2003; Daily Nation, as recently as December 6, 2007). This is generally attributed to degradation of marine resources and increase in number of fishers (Ochiewo 2004) but Mangi & Roberts (2007) specifically pointed to high levels of fishing effort coupled with the use of destructive gear. Although exact figures are not available, the general impression is that in recent years fishers have become more involved in economic activities outside fishing and that they spend more time on these activities. Fishers who cannot diversify their incomes, particularly during the *kusi* season (the low season), have to rely on relatives for financial aid. Many of these relatives are also under financial pressure and cannot offer assistance to others. At the same time, diversification is a two-way process. Due to declining employment opportunities and increasing pressure on land resources, more and more people are turning to fishing as an (extra) income source; many from coastal groups that have no tradition of sea fishing.

Table 5.1 Economic activities by study group (% households)

	Fishers (N=133)			*Non-fishers (N=80)*		
	Head	Wife	Others	Head	Wife	Others
Fishing	86.5	0.0	39.1	0.0	0.0	3.8
Farming	50.4	40.6	29.3	51.2	30.0	20.0
Self employment	14.3	18.0	17.3	52.5	18.8	21.3
Wage employment	3.8	0.8	21.8	33.8	1.3	25.0

Source: Household Survey

Little is known about household incomes and the income composition of local fishers. The impression is that both vary greatly among fishing villages and within villages but little is documented about whether fishers have other resources and, if so, the nature of these resources, and whether households are wholly or partly dependent on fishing. Household size is an important factor in this respect since it affects not only the household needs but also the available labour pool. The life phase of the household is also important in this respect (see Box 5.1 below). The economic activities of wives and other household members and their financial contributions are important, although it would be incorrect to assume that all income sources are pooled in one common household budget. The food security of fisher households, in turn, depends on the combination of available resources and livelihood strategies. The Household Survey in four locations (Map 3, p. 34) details the extent of economic activities, the household incomes and the income diversification of different groups.

Economic activities

The type and frequency of economic activities in households, including the participation of different household members in fishing, agriculture, wage employment and self-employment, are presented in Table 5.1. The last category consisted of activities, which, in practice, varied greatly in type and importance. Among the fisher households, nearly all heads were involved in fishing[16], about half were involved in farming, and 18% reported employment. Among the non-fishers, none of the heads was involved in fishing (by definition); nearly all were either self-employed or involved in wage employment. In this group, half the respondents were involved in farming.

Table 5.2 Farming characteristics by study group

	Fishers (N=133)	Non-fishers (N=80)
Farmland present (%)	64.7	73.7
Average farm size (acres)	3.7	5.6
Sale food crops (%)	3.8	11.3
Sale tree crops (%)	33.1	38.8
Sale milk (%)	6.8	3.8
Sale eggs (%)	–	2.5

Source: Household Survey

About three-quarters of the heads were married and about half of them had wives with an economic activity other than household chores. There was little difference between the fisher and non-fisher households in this respect. In 30-

Table 5.3 Responses to whether household income is sufficient to meet household needs (%)

	Fishers (N=124)	Non-fishers (N=72)
Sufficient	11.3	9.7
More than half needed	17.7	16.7
Half of what is needed	49.2	40.3
Less than half needed	21.8	33.3
	100	100

Source: Household Survey

Table 5.4 Household income composition (average; Ksh/week)

	Fishers (N=127)	Non-fishers (N=76)
Fishing	1439	0
Farming	305	262
Self employment	159	641
Wage employment	50	420
Total (s.d.)	1952 (1156)	1323 (854)

Source: Household Survey

Table 5.5 Income composition of fishers by activity diversification (average; Ksh/week)

	Single activity (N=54)	Multiple activity (N=57)
Fishing	1607	1356
Farming	25	283
Self employment	0	96
Wage employment	0	14
Total (s.d.)	1633 (1019)	1750 (1018)

Source: Household Survey

Table 5.6 Income composition of fisher households by earner diversification (average; Ksh/week)

	Single earner (N=68)	Multiple earner (N=43)
Fishing	1469	1548
Farming	157	324
Self employment	66	276
Wage employment	3	44
Total (s.d.)	1696 (1071)	2192 (989)

Source: Household Survey

40% of the households, wives were involved in farming; in about 20% of the households, they were involved in vegetable and food selling, plaiting *makuti* (coconut-leaf thatch) and other handicrafts.[17] Other adults in the household nearly always consisted of grown-up children.[18] In about 40% of the households, grown-up children, like their fathers, were involved in fishing, in 20-30% they were involved in farming and in about 40% in employment of some kind.[19] In all, fisher households reported economic activities of household members more frequently than non-fisher households, particularly of grown-up children. This was probably related to the larger household size of the former.

Apart from fishing, farming was the most common activity reported in both groups. Almost two-thirds of the fisher households had farmland (Table 5.2). The non-fishers reported farmland even more often and, perhaps more importantly, had plots that were larger in size than those of the fishers. The land was used to cultivate food crops that were mainly used for home consumption with little or none sold. Almost half the households cultivated tree crops and about a third sold part of their harvest. Approximately 10% owned cattle, 46% goats and/or sheep and 56% chicken and/or ducks. Sale of milk or eggs was negligible. In general, non-fishers had slightly more farm assets than fishers.

About 10% of the respondents stated that their income was sufficient to feed the household (Table 5.3) and 15-20% said that they earned more than half the income they needed. In many households, however, the respondents earned only half of what was needed and a quarter earned even less. The latter group of needy cases were more prevalent among non-fishers than among fishers (33% vs. 22%).

Household incomes

There was little difference between fishers and non-fishers in the number of households that received contributions from different household members. More than 89% of households had income from the head, 25-30% had income from wives, and about 20% reported income from 'other' household members. The total income of fisher households was estimated at Ksh 1,952 per week, al-

Box 5.1 Fisher livelihood strategies and income diversification

Reasons to become a fisher were the same in Takaungu and Uyombo, namely that there were few other jobs available. In Takaungu there was another option, although not very attractive, namely block cutting in the local quarry. 'Young' fishers often operated as migrant fishers; fishing elsewhere along the coast for periods of days or weeks. However, from the moment they acquired family responsibilities, they tended to become permanent at one landing site and, in turn, to teach 'new' fishers how to fish. Migration along the coast to fish elsewhere was quite common but it was found more often among fishers in Takaungu than in Uyombo.

Most fishers had another source of cash income besides that from fishing. One fisher household in Uyombo, which did not have an extra income, was assisted monthly by a brother of one of the household members. In Takaungu, one household reportedly had to buy on credit during the *kusi* season. Most households also had some agricultural land. One household in Uyombo only cultivated cash crops on the agricultural land; another was saving money to buy a *shamba*. A few households did not have land because "we do not need it". Most households with a *shamba* cultivated food crops and, in Takaungu, the households generally had greater food security than in Uyombo.

More than half the households produced cash crops. In Takaungu, few households sold products from coconut trees. In Uyombo, most households had coconut trees; a number of them sold *makuti* roofing during the *kusi* season. The main advantage of coconut trees was that one could receive cash money at any time of the year.

Households in Takaungu were more financially stable than in Uyombo. Most fisher households from Uyombo needed their cash income to meet household needs and to send their children to school. In Takaungu, the cash income was needed less in the short-term and, consequently, could be invested, for example, in the purchase of livestock.

In Uyombo, people were less able to help others and were more in need of help themselves: three households were assisted by people from outside the household and, in turn, two households assisted others. In Takaungu, one household was assisted from outside and four households assisted others. People from Uyombo requested help from relatives more often than in Takaungu. In Takaungu people generally expected to be able to rely on someone in times of need since they had assisted other people. This was not the case in Uyombo.

The life phase of the household was another important factor that affected household resources. A household with only young children had few people contributing to production (e.g. the father and the mother) and many people

(*continued on next page*)

Box 5.1 continued

consuming. Once the children grew up this pattern changed since the labour pool increased and the households could produce more than they consumed. When the parents became older and children moved out or were married and had children themselves, the income of the household decreased and consumption was higher than production again (Chayanov 1966). The households in Uyombo were more vulnerable during the first phase than in Takaungu.

In general, the most difficult periods were the first and last household phases. Fishers were aware of these problems and tried to avoid them in several ways. Arrangements were made to at least delay an unprofitable stage, for example, by keeping adult children at home. A fisher household could start more income generating activities, however limited they might be. To depend on social relations and count on assistance from a friend or relative was increasingly difficult since most people were facing the same problems.

Source: Versleijen (2001)

most 50% higher and significantly different than the Ksh 1,323 per week of non-fishers (Table 5.4).[20] Earnings from fishing comprised 74% of the income of fishers, with the rest coming from farm sales (16%) and 'employment' activities (10%). Income of non-fishers, in contrast, depended largely on self-employment (48%) or wage labour (32%) and was spread more evenly. Non-fishers had an average income from employment of Ksh 850 per week (vs. Ksh 464 for the fishers), although this offered only partial compensation. Income from agriculture was nearly the same as that of fishers and was relatively low, which agreed with earlier information on the extent of farming (see also Box 5.1).

From an income point of view, it was clearly unattractive to abandon fishing for land-based activities. This did not imply that diversification cannot be successful but that it can only be beneficial in combination with fishing. In conclusion, fisher households realized higher incomes than non-fisher households and the difference was mainly due to earnings from fishing.

With an average household size of 7.7 people, the household income of Ksh 1,952 per week (mentioned earlier) amounted to a monthly average of Ksh 1,090 per person, below the national poverty line of Ksh 1,239.[14] From these figures it was calculated that 59% of fisher households fell below the poverty line. This figure is smaller than the 64% reported among the rural population in Coast Province overall[21] and it is evident, therefore, that fishers do not fare worse than the province's rural population in general.

Income diversification

Diversification among fishers occurred when the household was not dependent on only one income source. This lofty position was achieved in two ways. The first occurred at the individual level, when fishers had more than one economic activity which is referred to as 'activity' diversification. The second occurred at the household level, when other household members had incomes that contributed to household expenses, and is referred to as 'earner' diversification. Activity diversification was more common than earner diversification (51% vs. 39%) and the two types occurred largely independent of each other.[22]

To address the question whether diversification resulted in changes in income level or changes in income composition, the 'single' and 'multiple' livelihood scenarios were examined among 111 fisher households (see Chapter 2). In the case of activity diversification of the head of a household (Table 5.5), his income from fishing decreased by Ksh 250 per week, but this was compensated by income from farming and self-employment, so that the head's total income was slightly, but not significantly, higher. Activity diversification led to a greater spread of income, but not to substantially higher incomes for the group as a whole. This was different for earner diversification (Table 5.6) where incomes from fishing were nearly the same among households of single and multiple earners. Among multiple earners, farming and self-employment were the main sources of extra income. Earner diversification led to higher incomes without any shift away from fishing.

Table 5.7 presents the distribution of the two types of diversification in fisher households. In 19% of households, diversification of both kinds was present, whereas no diversification was found in 29% of the households. Furthermore, 32% and 20% reported either activity or earner diversification, respectively. Households with both kinds of diversification had the highest average household income, followed by households with earner diversification only, then activity diversification only and finally no diversification. Analysis of variance confirmed that there were significant income differences because of earner diversification but not because of activity diversification. These differences were independent of household size.[23]

Table 5.7 Household income by type of diversification (average; Ksh per week; fisher households)

	No diversification	Activity diversification	Earner diversification	Double diversification	Total
No. cases	32 (29%)	36 (32%)	22 (20%)	21 (19%)	111 (100%)
Household income	1670	1718	2102	2286	1888

Source: Household Survey

In conclusion, earner diversification resulted in higher incomes and activity diversification added little to income levels for the group as a whole although the trends were not the same when boat captains (and independent fishers) and crewmembers were examined separately (next section).

More about activity diversification

There was little difference between the two groups in terms of total income although boat captains earned 91% of their incomes from fishing and crewmembers 79% (Table 5.8). A boat captain was either the owner of a boat or leased a boat from a *tajiri* and, as such, received an extra share of the catch. Crewmembers, however, had higher incomes from farming and self-employment and thus managed to earn the same total income.

The income composition of boat captains and crewmembers with different livelihood strategies is presented in Table 5.8. Boat captains whose only activity was fishing had the highest income from fishing, about Ksh 1,750 while captains who had other economic activities averaged Ksh 1,350 per week from fishing. Among crewmembers, the income from fishing was also about Ksh 1,350 per week, irrespective of whether they had other economic activities or not. The picture changed, however, when the head's total incomes were examined. These were not easy to interpret because the trends differed within the sub-groups. Among boat captains, income from a single livelihood was higher than that of a diversified livelihood. The opposite was true for crewmembers. Boat captains had to invest time in the repair and maintenance of equipment to be successful and needed time to organize fishing trips and make arrangements for the sale of the catch. These preparations were vital to the success of the fishing enterprise. If time was spent on other economic activities, it is likely that income from fishing would decrease accordingly. This was not the case for the crew. Diversified crew managed to use the remaining time economically, adding more than 50% to their fishing income with non-fishing activities and realized the highest income of all groups. Crewmembers without extra economic activities, however, were in the lowest income group.

Table 5.8 Income composition of fishers by fisher status and activity diversification (average; Ksh/week)

	Boat captains			*Crewmembers*		
	Total (N=73)	Single activity (N=33)	Multiple activity (N=40)	Total (N=38)	Single activity (N=21)	Multiple activity (N=17)
Fishing	1559	1744	1350	1357	1352	1362
Farming	135	0	287	247	33	511
Self-employment	20	0	43	118	0	265
Wage labour	0	0	0	5	0	12
Total income head	1714	1744	1680	1727	1386	2149
(s.d.)	(934)	(1046)	(867)	(1067)	(770)	(1245)

Source: Household Survey

Food consumption

The success of livelihood strategies is not only reflected in incomes and income diversification but is also shown by the food security of households. General information on food habits and food consumption along the Kenyan Coast and among the study population is presented in Appendix 3 (p. 123) together with a description of the method of data collection. The text below focuses on possible differences in food security between fishers and non-fishers and the possible relations with household diversification. Three aspects of food security were studied: food production, food stocks and food consumption. We examined: (i) the percentage of households that grew or produced one or more foods in the three main food groups (staple foods, legumes/greens, animal products); (ii) the stock of maize/cassava and the estimated size of the stock, expressed as the number of months it is expected to last; and (iii) the number of days per week that the foods were prepared, aggregated for each of the three food groups.

Starting with subsistence food production, non-fishers produced staple foods (maize and cassava) and legumes/greens more often than fishers (Table 5.9). In contrast, fishers produced more animal products, notably fish from their own catches, than non-fishers.

The percentage of households with stock of locally grown maize and cassava was larger among non-fishers (~70%) than among fishers (~60%). The average food stock of non-fishers was 4.6 months and of fishers was 4.2 months (Table 5.9).

Table 5.9 Food security characteristics by study group

	Fishers (N=133)	Non-fishers (N=80)	p
Food production (%) [1]			
Staple foods	56	68	.09 [5]
Legumes & greens	56	64	.24 [5]
Animal products	96	54	.00 [5]
Food stock			
H'holds with food stock (%) [2]	56	69	.06 [5]
Size of food stock (ave.) [3]	4.2	4.6	.61 [6]
Food consumption (ave.) [4]			
Staple foods	10.0	10.5	.15 [6]
Legumes & greens	7.4	8.1	.24 [6]
Animal products	7.7	6.2	.00 [6]

Source: Household Survey
(1) Households with home production of one or more of foods mentioned
(2) Households reporting stock of staple foods
(3) Estimated duration of stock of staple food (number of months)
(4) No. of times that foodstuffs are eaten in course of one week
(5) Chi Square
(6) ANOVA (Main effects for fisher status)

Table 5.10 Food security of fisher households by earner diversification

	Single earner (N=68)	Multiple earner (N=43)	p
Food production (%) [1]			
Staple foods	46	58	.20 [5]
Legumes & greens	46	56	.29 [5]
Animal products	96	100	–
Food stock			
Households with stock (%) [2]	46	58	.20 [5]
Size of food stock (ave.) [3]	3.3	4.6	.09 [6]
Food consumption (Av.) [4]			
Staple foods	10.3	10.0	.45 [6]
Legumes & greens	6.3	8.0	.03 [6]
Animal products	7.9	7.8	.72 [6]

Table 5.11 Food security of fisher households by activity diversification

	Single activity (N=54)	Multiple activity (N=57)	p
Food production (%) [1]			
Staple foods	17	83	.00 [5]
Legumes & greens	13	84	.00 [5]
Animal products	96	98	–
Food stock			
Households with stock (%) [2]	17	83	.00 [5]
Size of food stock (ave.) [3]	1.0	6.4	.00 [7]
Food consumption (Av.) [4]			
Staple foods	10.3	10.0	.57 [7]
Legumes & greens	5.3	8.6	.00 [7]
Animal products	8.1	7.7	.27 [7]

Source: Household Survey
1-2-3-4 See Table 5.9
5 Chi Square
6 ANOVA (Main effects for earner diversification; Covariate activity diversification)
7 ANOVA (Main effects for activity diversification; Covariate earner diversification)

Similarly, the frequency of preparation of foods from the three respective food groups also showed differences. Fishers reported slightly fewer staple dishes and fewer legumes/greens and non-fishers consumed cassava and greens more frequently. From other research, we knew that the consumption of animal products tended to be small and that staple foods and grain legumes formed the bulk of calories and proteins in the coastal diet (Klaver & Mwadime 2000).

However, fisher households were an exception as they consumed animal products more often than non-fishers, mainly because of higher fish consumption. Still, it is noteworthy that non-fishers ate fish more often than other rural households not living near the coastline, as reported by Hoorweg *et al.* (1995).

Altogether, the differences in food security between fishers and non-fishers were small but fit with the livelihood patterns of the groups. Differences were related to two factors. Non-fishers had more farming assets and, in line with this, had more food stocks and ate more green vegetables. Fishers, in general, kept part of their catches for home consumption and ate fish more often than non-fishers.

Further analyses examined the food security in households with 'earner' diversification and households with 'activity' diversification. Results, including the percentage-wise figures and statistical tests, are listed in Tables 5.10 and 5.11.

❖ Food production. Households with either type of income diversification scored higher in terms of production of staple foods and legumes-greens (the score for animal products was near to 100% in all groups). Households with earner diversification scored roughly 10% higher than single earner households. Among households where the head had 'multiple' activities, about 85% cultivated food crops. When the head was exclusively occupied with fishing, this figure was only 15%, a highly significant difference.

❖ Food stocks. In households with multiple earners, almost 60% reported food stock. Calculated over all households, the stock was expected to last for a mean of 4.6 months (this figure would be 7.9 months when considering only the households that reported stock). Households with one income earner had less stock than multiple earners, 46% of the households and 3.3 months duration, respectively although the differences between multiple and single earner households were not significant. Differences due to 'activity' diversification were larger. When the head of a household had one income (from fishing), only 17% of the households had any stock of staple foods, which was expected to last only one month. In contrast, when heads had more than one income source, 83% of the households had staple foods in stock and the stock was estimated to last more than six months.

❖ Food consumption. There was no difference in the number of times that staple foods were prepared during the week; the figure was high in all groups with an average of nearly 1.5 times a day (Table 5.10 and 5.11). Significant differences did occur in the frequency with which legumes/greens were consumed in households with more than one income earner as well as households where the head had multiple activities. Fish dishes were consumed more often in households where the head restricted himself to fishing as the only activity.

The effect of diversification on food practices among fishers can be summarized as follows: diversified households relied more on food consumption from own production, had more stock of staple foods, and consumed greens/legumes more often. In general, results presented a consistent picture that pointed at better food security among fisher households with activity diversification, more so than with earner diversification. It is possible that households with earner diversification relied more on food purchases because of higher household income.

Conclusions

This chapter addresses the issue of fisher incomes and household livelihoods by examining fisher and non-fisher households living in the same vicinity. Since the households were paired, there was no difference between the two groups in regards to the external variables of agricultural potential of the area and (distance to) employment opportunities. Fisher households were engaged in more activities than non-fisher households and had the advantage of a broader resource base. Non-fishers had more farming assets but fishers also engaged in farming. Income figures supported the above observations. Fisher households earned 50% more income than non-fishers, a difference that was due mainly to earnings from fishing. Fishing was vital to the income of fisher households.

Cinner & McClanahan (2006) also reported that fishing was a primary occupation for many households, rather than a supplementary income source. Judging from the comparison with non-fishers, it was difficult to compensate for fishing income with other economic activities. Although non-fisher incomes were more evenly spread over different activities, their incomes were considerably lower.

The small differences in food security found between fishers and non-fishers were not to the clear advantage or disadvantage of either group. Non-fishers farmed more often, had more food stocks and consumed green vegetables more often. Fishers generally kept some of their catch to take home and ate fish more often than non-fishers. Among the fishers, diversified households relied more on own production and had more staple foods and greens/legumes. This was particularly so in households where heads reported activity diversification. Food security in households with activity diversification was better than in households with earner diversification, although the latter had higher incomes to purchase foodstuffs.

Despite the relatively favourable position of fishers in their local communities, it is still possible for these communities near the coast to be poor with income levels below those in rural areas elsewhere in Coast Province. This has, however, been shown not to be the case: the incidence of poverty among fishers was not higher than among the general rural population in Coast Province. Therefore, the view that fishers are destitute, among the 'poorest of the poor', are trapped in a hopeless situation and are desperate for other opportunities did not apply here. The fishing sector was not an employer of last resort and income diversification was not a means to escape fishing but rather it was an integral part of livelihood strategies. The other side of the coin is that non-fishers will be attracted to fishing in order to improve their livelihoods, and this was indeed happening with the entry into the sector of many Mijikenda fishers. Nevertheless, it must be realized that many fisher households lived below the poverty line and, as such, were part of the region's general poverty problem.

Earner diversification resulted in substantially higher household incomes. Activity diversification added substantially to the incomes of crewmembers but not to the income of boat captains. Few fisher households lived without diversification of any kind, which confirmed the importance of this income strategy. Earner and activity diversification can be regarded as different strategies, each with its own advantages and disadvantages. Activity diversification means that the head can decide how the income is used, but the income is limited by what that one person can earn. It is reasonable to assume that the frequency of activity diversification decreases with age whereas the frequency of earner diversification increases with age. Wives can start to earn an income when the older children are able to care for their younger siblings. At a later stage, children can also earn money. Earner diversification means that the household has more income but the head does not have to put in any more work. The disadvantage from his point of view is that he does not have sole control of the income. Still, it is likely that the wife will spend most or all her earnings on household necessities. As for older children, they will probably get married and, therefore, contribute earnings only for a limited number of years.

6

Marine conservation

with Nicole Versleijen

The coastal and marine environments of Kenya are threatened by naturally occurring processes, growing subsistence needs of the coastal population, and increased economic activities in general (Hoorweg 1998). Examples of natural processes are coral bleaching, sea level change and beach erosion. Growing subsistence needs are behind the overharvesting of mangrove trees, illegal shell collection and intensive fishing. Growing economic activities result in increases in sewage and waste disposal from tourist hotels, industrial pollution of waters, and siltation at river exits as a result of soil erosion upcountry. The first national environment plan in 1994 listed many of these issues but efforts at 'integrated coastal management' since then have been limited to the Mombasa and Diani areas and, according to some observers, have focused more on infrastructural development and resource access than resource management and protecting biodiversity (McClanahan, Mwaguni & Muthiga 2005c).

Artisanal fishers can also contribute to the degradation of marine resources (Payne 2000). Intensive fishing can affect the ecological balance, for example, causing a rise in sea urchin abundance and a resulting loss of benthic cover (McClanahan & Obura 1995). Destructive fishing practices, such as the use of beach seines and explosives, can alter the terrain and affect the ecological balance of the reef and seafloor. Local fishers generally do not approve of destructive fishing methods since they are aware that this will ultimately lead to poorer catches. Indeed, nearly all fishers were concerned with the degradation of marine resources and mentioned declining fish catches (Glaesel 2000). Other reasons given by fishers included increased number of fishers, gazettement of no-take-areas, rough weather (e.g. the El-Niño of 1997/98) and competing fisheries such as commercial trawling.

Modern methods of marine conservation attempt to minimize the impact of intensive fishing in a number of ways, in particular by (i) limiting the number of fishers, (ii) restricting access to fishing grounds, and controlling (iii) type of gear and (iv) frequency of fishing. This chapter explores these features before choosing a 'key' indicator for each attribute, subsequently used to test relations with income diversification.

Fisher number

In 1999, all that was required to fish on the Kenyan Coast was a fishing license from the local Fisheries Office, at 100 shillings for a one-year period (about $1.25 at the time). However, controls were lax and many fishers did not purchase a license. At the start of research, the official count by the Department of Fisheries (1996-1998) was 1,000 fishers along the combined Malindi and Kilifi coasts. The respondents at the five coastal tracts in the Fisher Survey, however, estimated a much higher number, 1,800 fishers. The largest numbers of fishers were reported for Ngomeni (~400), Malindi (~500) and Mida (~350) while in Kilifi (~330) and Takaungu (~230), further south, the numbers were somewhat smaller. The total number of 1,800 had to be increased for landing sites that were not covered in the survey, as well as for other smaller landing sites on this stretch of coast of 125 km, arriving at a rough estimate of 2,500 to 3,000 fishers. This is almost triple the official figure.[24] Extrapolation of these figures to the full length of the coast, 600 km from Kiunga to Vanga, would arrive at 10-12,000 fishers in all.

A fisher committee, headed by a chairman, existed at most landing sites. New fishers at Takaungu, for example, had to pass through the chairman to obtain permission to fish. Reasons to deny permission were mainly the type of gear used by the fisher and his reputation. However, the chairman himself admitted that not all fishers at Takaungu had his official approval. But, as long as the fishers did not use destructive gear and did not cause problems, they were allowed to carry on. A village committee existed in Uyombo, with fishers not living in the direct vicinity of the landing site being members as well. Here, also, new fishers required approval from the committee. The role of this committee was mainly to deal with complaints, facilitate internal communication and represent fishers in talks with external parties (For example, discussions with KWS concerning the Watamu Marine Park, which were taking place during the period of research).

Nearly all boat captains and independent fishers (N=83) in the Household Survey (91%) were negative about current fishing trends and reported declining fish catches. An increased number of fishers was mentioned most frequently as the cause. If there was indeed a major decline in catches and incomes, why did people start fishing, or allow other people to start fishing? The answers mostly referred to easy access and lack of alternative employment. Fishers believed that anyone can fish whenever he wants and in the way he wants.

> *Everybody can start fishing whenever he wants and in the way he wants. It is not like you have to look for it a long time and to go through a lot of trouble.* (Mijikenda fisher, Takaungu)
>
> *If there were other jobs I would do something else, but you know it is hard to find a job these days, even the tourist hotels are not offering many jobs anymore.* (Mijikenda fisher, Takaungu)
>
> *My family had been farming for a long time, my grandfather and his father and so on. But when I was young, the harvest was not that good anymore and it would become a problem for me to live from farming alone when I wanted to start a family. So I started fishing. Other fishers took me out and taught me how to do it. And some of my sons started to help me fishing and they will become fishers as well!* (Mijikenda Fisher, Uyombo)
>
> *We do not own the sea, it is the KWS who thinks you can own sea! Sea is for everybody; so one fisher can never deny another fisher to go fishing. Unless that fisher must be fishing in a way that is not accepted by the fishers. You know like the Wapemba, we chased them because they were ruining everything!* (Former Bajun fisher, Uyombo)

Asked about their willingness to stop fishing, 87% of fishers responded positively. This was an unexpected high percentage but, also, somewhat deceptive because old age was mentioned as the foremost (and inevitable) reason to retire from fishing (71%). But, it is noteworthy that 54% of fishers was willing to take alternative employment, if available, although it is unlikely that they would stop fishing completely if the opportunity occurred. Only 1.2% mentioned low catches as a reason to stop (Table 6.1).

Table 6.1 Fishing trends and reasons to give up fishing*

	All
Fishing Trends (%)	
decreasing catches	91.3
Reasons to give up fishing (%)	
old age	71.1
employment opportunities	54.2
low catches	1.2

Table 6.2 Fishing ground restrictions by season (%)*

	Season	
	High	*Low*
Avoid		
lagoon/inshore	72.3	28.9
reef	30.1	34.9
outreef/deep water	41.0	85.5

* Household Survey, N=83. MR

Willingness to stop fishing was related to age and fishing income. Fishers over 40 years mentioned 'age' more often as a reason than younger fishers (91% vs. 56%). Younger fishers were more willing to try other employment than older fishers, who possibly saw fewer opportunities open to them (Table 6.3). Also, fishers with a low fishing income were less willing to exchange fishing for other employment. Perhaps they were realistic enough to know that, at best, they could attain unattractive, menial jobs.

Fishing grounds

Fishers frequented one or more types of marine habitats during their fishing trips. These included the lagoon, inshore grounds, the reef itself, and grounds beyond the reef. Many also ventured into deep waters, outside the protection of the reef, where they were more exposed to the sea and foul weather. For most fishers, the deep waters were second or third choice because of inadequate vessels. Regular deep-sea fishing was the domain of the larger, sturdier vessels as well as the sport fishers and commercial fleets.

Alternatively, it is possible to examine the fishing grounds that are avoided at certain times of the year. In this respect, there was a clear difference between the high and low seasons (Table 6.2). During the *kusi* season, when the sea can be rough, fishers avoided the out-reef areas and deep waters. During the *kaskazi* season, they fished the lagoon and inshore areas less often giving these grounds some respite. Fishers from landing sites nearby often mentioned the Marine Parks as no-go areas (80%). Artisanal fishers were aware of the important role of the reef where many species spawn and breed. However, pressure on the reef was more or less the same during the two seasons, as it was mentioned by about 60% of the fishers in each season.

Table 6.3
Willingness to stop fishing for alternative employment by selected variables (%)*

	Age group			*Fishing income*	
	young[1]	*old*[2]		*low*[3]	*high*[4]
Yes, willing to stop	65.1	34.4	Yes, willing to stop	31.6	58.9
No, not willing	34.9	65.6	No, not willing	68.4	41.1
	X2: p=.008			X2: p=.04	

* Household Survey, N=83.
(1) <40 years, N=43; (2) ≥ 40 years, N=32;
(3) < Ksh 1,000/week, N=19; (4) ≥ Ksh 1,000/week, N=56.

Restricting access to fishing grounds, in the form of a seasonal or all-year ban, is an important conservation measure. In the past, there were restrictions such as the *sadaka*, traditional ceremonies in which certain areas were designated as off-limits for local fishers, but these ceremonies have fallen largely into abeyance. In November 2000, a *sadaka* was called in Takaungu but only nine fishers attended. The nine were all Muslims and over the age of forty. The ceremony itself consisted of eating on the beach, offering some food to the sea and not fishing in that spot on the day of the ceremony.

> *We used to prepare food and invite other fishers, take the food to the beach and eat all together and go home, the leftover food is given to the sea.* (Swahili fisher, Takaungu)

> *When there is a high catch, the fishers gather at the beach and roast and eat the fish all together, but this is not anymore.* (Mijikenda fisher, Takaungu)
>
> *Some Mavumba (pounded fish which has a very strong smell, the smell is the important thing of it, it can be rotten fish as well) are taken to the sea and some words are said and celebrations are done. This can be anywhere in the sea, the place is chosen by all the fishers together.* (Mijikenda fisher, Takaungu)

Views differed considerably on how the ceremonies should be conducted and only participants in the ceremony actually refrained from fishing in the designated area. The reasons given for participating in the ceremony were that: (i) they used to have *sadaka ceremonies* and this was enough reason to continue; and (ii) the Gods had to be pleased for the fishers to improve their catch.

> *Before conducting the ceremony, the gods have to be pleased. The elder fishers have to speak some words and then some rice and fish have to be given to the sea. They used to conduct the ceremony, but now three years have passed without conducting the ceremony, it is like people care less about it.* (Bajun fisher, Takaungu)
>
> *The ceremony used to be conducted every year to please the gods, regardless of a high or a low catch. In those days most fishers were Muslims and they all agreed that the ceremony should be performed; nowadays however there are a lot of non-Muslim fishers. There is no co-operation between the fishers anymore and the non-Muslim fishers are afraid that when they conduct the ceremony a few days afterwards a non-Muslim might drown in the sea.* (Swahili fisher, Takaungu)
>
> *There is a ceremony in which blood should be given to the sea. A goat is slaughtered and prepared and eaten. Some is given to the sea. Elder fishers say some words to the gods of the sea to ask them for a higher catch. After the ceremony there should not be fishing at the spot of the ceremony for a week. This ceremony is not there anymore, the fishers have become too many and are not co-operating anymore. The elder fishers who were always arranging this have died years ago. I think the last ceremony like this must have been 10 years ago.* (Swahili fisher, Takaungu)

The main restriction in force nowadays is that of the Marine Protected Areas which consist of Marine National Parks and Marine National Reserves (see Chapter 1). A Marine Park is an area where neither fishing activities nor plant or animal collection are allowed. Fauna and flora are fully protected inside the parks. In Marine Reserves, fishing by artisanal fishers is allowed, but is restricted by regulations stipulated in the Fisheries Act (Kenya 1991). Appendix 4 (p. 127) presents a list of the existing regulations in the Marine National Parks and Reserves. The Marine Protected Areas were managed by KWS and wardens patrolled regularly. Fishers were allowed to pass through the Parks with their vessels to reach their fishing grounds in the Reserves. Fishers in unprotected areas were expected to keep to the general fishing regulations, but there was little inspection.

Marine Protected Areas offered advantages as well as disadvantages for the fishers living nearby (see also Box 6.1). The main disadvantage was that parts of the traditional fishing grounds were off-limits. Often, Parks were established on what the fishers considered good fishing grounds, notably the breeding and spawning areas of many fish species (Glaesel 1997a; Ochiewo 2004). Other areas had less to offer in way of marine resources and, besides, were often more difficult to reach. A positive effect that was expected from the fishing restrictions was an increase in species diversity within the Parks. A second effect was an increase in fish numbers, which should spill over into the Reserves and surrounding areas to the benefit of local fishers (Kaunda-Arare & Rose 2004a, 2004b; Roberts *et al.* 2001). However, this effect can be nullified by a greater concentration of fishers in a smaller area (Ochiewo 2004). The large majority of fishers from landing sites near a Marine Park indeed avoided these grounds. (In Mayungu and Uyombo, 80% of the fishers mentioned the Parks as off-limits both in the high and the low seasons). At the same time, almost three-quarters of the fishers at the Watamu landing site listed the Watamu Park as one of the main problems with which they had to cope. In fact, fishers in Uyombo showed considerable resentment against the Park as well as the wardens of KWS (Versleijen & Hoorweg 2006). Another study also concluded that fishers did not think that they benefited from 'area management' (McClanahan, Davies & Maina 2005a).

Fishing gear

Fishers were flexible in their use of gear, although they usually had strong preferences based on experience and the expected catches (Tunje & Hoorweg

Box 6.1 Resource conservation as a challenge facing artisanal fishers

Attitudes towards conservation differed considerably among fishers. In Takaungu, most of the (younger) fishers admitted that there was a need for conservation. They were also aware that whatever form the conservation measures would take, this would affect their fishing practices. The alternative, they envisioned, was to offer people other employment. However, given the poor employment opportunities, rather the opposite was expected to occur. People who had no other income would turn to fishing and, consequently, an even higher pressure on marine resources would result. In Uyombo, people were generally adverse to conservation measures. According to most fishers, their catches were low since there were too many fishers in a small area, which was limited by the Marine Park as well as by rough seas. The downward spiral of declining fish stock - less income - more school drop outs - more fishers - more exploitation of fish stocks was quite strong in Uyombo but was also present in Takaungu.

Fishers in Uyombo generally had a negative attitude towards the nearby Marine Park. They claimed that fish stocks were still declining after the gazettement of the Park. They also claimed that they had not benefited from the presence of the Park and they suggested various ways to correct this. Firstly, parts of the Park should be opened for fishing during the *kusi* season. Secondly, the fishers should receive parts of the gate collections of the KWS. Thirdly, the Marine Park generated jobs such as rangers, hotel employees, safari guides, beach operators (curio sellers) and boat operators, but these were often not available to the fishers, since they lacked sufficient training and/or starting capital. Therefore, benefits for the fishers were few and this has resulted in considerable anger and aggression towards the Marine Park. The danger is that the existence of the Park can make fishers adverse to conservation measures in general because they equate conservation with low catches.

To understand the adverse attitude of fishers towards the Watamu Park, it was necessary to realize that the interests of the KWS and the fishers were almost conflicting. Whereas the fishers wanted to land a large catch and improve their income, KWS wanted to control and limit fishing activities. Clearly, the attitudes towards conservation were affected by the presence of the Marine Park and more positive attitudes existed in Takaungu than in Uyombo. In Uyombo, many fishers had abandoned the idea of conservation, claiming that it was an idea of the *mzungu* (white man) and the government who only wanted it for their own benefits. As a result, they were probably less likely to participate in conservation programmes.

Source: Versleijen (2001)

2003). Many fishers reported two or more kinds of gear; only 30% of fishers limited themselves to one specific gear (Table 6.4). Gear differed greatly in their effect on the environment, some being potentially damaging, others not.

Roughly, there were three types of destructive effects: (i) damage to marine environment; (ii) capture of non-targeted species; and (iii) capture of immature targeted species. Not only the type of gear but also the area where the gear was used, and the way it was used, determined whether the gear indeed did damage or not. Traditional gear were generally less harmful than modern gear but the former were on the decline.

Traditional gear included traps, fences, spear guns and poison. The portable fish traps were fairly light and were used on the reef without adverse effects. Spear guns were considered destructive to the corals. Although the method was not damaging in itself, fishers sometimes used a long metallic rod *(mkonjo)* to break corals where fish took refuge. In addition, spear guns damaged the coral when fishers missed their target. Traditional fish poison was destructive not only to the fishery resources, but also to other organisms such as birds that eat dead fish.[25]

Table 6.4 Fishing gear characteristics (%)*

Fishing gear (%)*		*Net mesh size* (%)**		*Use destructive gear* (%)*	
nets	73.4	< 1.0 inch	3.4	beach seine	}
lines	62.8	1.0-2.5 inch	77.4	net mesh < 1.0'	} 15.6
traditional	9.0	3.0-4.5 inch	72.6	spear gun	}
other	9.5	> 5.0 inch	31.5		

* Fisher Survey, N=199, MR ** Fisher Survey, N=146, MR

Modern gear included nets and lines in almost equal proportion. The use of a gill net was destructive if it entailed trampling on the reef crest by fishers. However, when used in areas where corals were absent, gill nets rarely caused damage, although small fish became entangled in the nets. Small-mesh nets were destructive because they caught many young and immature fish as by-catch. Beach seines, moreover, were dragged along the seabed, overturning the sea bottom and damaging underwater vegetation. Explosives (*baruti*) not only killed fish and other marine life indiscriminately but also damaged the habitat.

They often reduced the reef to a layer of small pieces of coral and loose rubble.[26] In contrast, fishing lines were not considered destructive when used without overturning corals.

Fishing vessels and fishing gear differed considerably among coastal tracts because of differences in local marine conditions and the abundance of local fish species. The most popular gear was the (gill) net with mesh sizes between 2.0 and 4.5 inches. More than half the fishers used lines (25% reported long lines). Traditional gear were reported by fewer than 10% (Table 6.4).

About 15% of fishers freely admitted to using destructive gear – 9% reported using spear guns, 5% mentioned beach seines and 3% used a net mesh size of less than 1 inch. These gear were used more often by Mijikenda than Bajun fishers.[27] There was no consistent relation with age although spear guns were found slightly more often among younger fishers.

Fishing frequency

A final factor affecting the pressure on the marine environment was the frequency of fishing. Generally fishers set out five to six days a week and rested one or two days. Fridays were non-fishing days for many (57%) while others choose not to fish on other days of the week. Reasons to take a day off included religious observance, time for family, maintenance of gear and craft and avoidance of high tides and rough waters.[28]

Most respondents fished once a day for about four hours. Forty percent of the fishers went out six times a week, with a large variation among the other 60%. About a third of the fishers reported eight or more trips a week and, therefore, they either went out more than once a day or combined day and night fishing.[29] This was found, in particular, among fishers in Takaungu.

Table 6.5 provides further information on the fishing frequency during the high and low seasons, notably the duration of the fishing season and the number of trips per week. The duration of the high season averaged about 5.5 months and that of the low season almost 4.0 months, which left about 2.5 months without fishing activities. Many fishers did not go out during the height of the *kusi* season. The frequency of fishing trips differed slightly with 8.2 trips per week in the high season and 7.2 trips per week in the low season. The average number of annual fishing trips was estimated at 315 trips although the variation was considerable. About 25% of fishers made an estimated 360 trips or more while 25% made 210 trips or less.

Table 6.5
Fishing frequency by season (av/s.d.)*

	High	*Low*
Duration season	5.4	3.9
no. months	(1.9)	(1.2)
Fishing frequency	8.2	7.2
no. trips/week	(2.9)	(2.5)
Fishing frequency	193.1	121.1
no. trips/season	(99.8)	(58.9)

Table 6.6
Fishing frequency by age group (%)*

	Young[1]	*Old*[2]
low frequency[3]	43.5	64.3
high frequency[4]	56.5	35.7
	X2: p=.006	

* Fisher Survey, N=199
(1) <40 years, N=124; (2) ≥40 years, N=70;
(3) <300 trips/yr, N=99; (4) ≥300 trips/yr, N=95.

Frequency of fishing was related primarily to the type of vessel; motorboats generally went out more often than non-motorized vessels. Furthermore, fishers with large vessels (*jahazi* and *mashua*) went out more often than fishers with small vessels (*dau* and different canoes), particularly during the *kusi* season. Age of fishers was also a factor as younger fishers went out more often than older fishers (Table 6.6).

Income diversification and fishing practices

Thus far, we have identified four elements of fishing activities that can place pressure on the resource: number of fishers, fishing grounds, fishing gear and fishing frequency. For each element, a key indicator was identified and validated, namely (i) willingness to stop fishing for alternative employment; (ii) frequenting the lagoon and inshore grounds; (iii) use of destructive gear; and (iv) annual number of fishing trips. Earlier (Chapter 5), the analysis resulted in two kinds of income diversification, namely, earner diversification and activity diversification. Earner diversification means that there is more than one income earner in the household whereas activity diversification means that the fisher has more than one economic activity and is not dependent only on a fishing income. The question to be addressed is whether income diversification resulted in less pressure on marine resources. If a relation exists between diversification and fishing practices, it is expected to show firstly with activity diversification since this affects fisher incomes more directly. An eventual relation between earner diversification and fishing practices would be expected to be less pronounced.

Table 6.7 presents the results for earner diversification, comparing fishers who are single earners with fishers with more than one income earner in their households. No significant relation was found with fishing practices. In general, income in rural households is not pooled and, thus, offered the fishers little incentive to alter their dependence on fishing and change their fishing practices. In addition, the extra income was needed for the generally larger households and was not under the direct control of the fishers.

Table 6.7
Fishing practices by earner diversification

	Single earner	*Mult. earners*
A. Fisher number (%)**		
Willingness to stop fishing for alternative employment	57.8	45.5
	X2: p=.28	
B. Fishing grounds (%)*		
Frequent lagoon and/or inshore grounds	43.4	42.5
	X2: p=.92	
C. Fishing gear (%)*		
Use of damaging gear	17.6	7.5
	X2: p=.12	
D. Fishing frequency (%)*		
Number of annual trips above average	49.0	48.7
	X2: p=.97	
* Fisher Survey	N=159	N=40
** Household Survey	N=45	N=33

Table 6.8
Fishing practices by activity diversification

	Single activity	*Mult. activities*
A. Fisher number (%)**		
Willingness to stop fishing for alternative employment	58.8	51.2
	X2: p=.50	
B. Fishing grounds (%) *		
Frequent lagoon and/or inshore grounds	33.3	57.3
	X2: p=.00	
C. Fishing gear (%) *		
Use of damaging gear	8.5	25.6
	X2: p=.00	
D. Fishing frequency (%)*		
Number of annual trips above average	50.0	47.5
	X2: p=.73	
* Fisher Survey	N=117	N=82
** Household Survey	N=34	N=43

Fishing practices, however, did show a relation with activity diversification although, in contrast to our expectations, the effects were harmful (Table 6.8). Two of the indicators differed significantly between fishers with a single economic activity and fishers with multiple activities. The fishers with multiple activities mentioned lagoon-inshore more often as their fishing grounds. Apparently, these fishers went on fewer (longer) trips outside the reef, presumably because they had work commitments onshore.[30] Fishers with multiple activities also reported the use of destructive gear significantly more often; it is likely that they had less time to fish than needed for regular boat fishing.

The two remaining indicators showed no significant differences. However, fishers with multiple activities were slightly less willing to stop fishing for alternative employment than fishers with single activities (51.2% and 58.8%, respectively). The former already had alternative employment, which could be combined with fishing. Although this was not a statistically significant trend, the opposite was certainly not the case, in that there was no indication that fishers with multiple incomes were more willing to exchange fishing for other employment.

Conclusions

This chapter offers an assessment of the pressures that fishers exerted on the marine environment and how these were related to the existing livelihoods of fisher households. Most fishers in the study area were aware of the degradation of marine resources and mentioned declining fish catches. They attributed this mainly to an increased number of fishers. This feature and three other elements of fishing activities that affected the marine environment were examined, namely fishing grounds, type of gear, and frequency of fishing.

The number of fishers has been increasing over the past decades with the entry of many Mijikenda into the arena, a group not known for its fishing in the past. The reasons for their entry were the open and easy access of the resource, the lax enforcement of license regulations and the need for employment. Half the fishers expressed an interest to opt for alternative employment, if available, although it is doubtful whether they would abandon fishing completely if they did find employment. It is more likely that they would try to combine the two, as did many of the new entrants. Fishers with a low fishing income were less willing to choose alternative employment, which is in line with the finding that families with higher incomes are usually in a better position to diversify than poorer families (Ellis 1999).

The majority of fishers used nets with approved mesh sizes and lines (including long lines). Traditional gear have become less popular with time and were used by only 10% of fishers. Generally, methods which involve walking or standing on the shallow reef crest, overturning boulders and dragging gear over the reef or sea bottom were destructive. This led to loss of diversity of the benthic substrata, resulting in fewer places for concealment and less habitat diversity for fish species. Gear that were destructive included spear guns, beach seines and other nets with small mesh sizes. Nets with small mesh sizes were particularly destructive as they capture non-targeted species and juveniles of the targeted species. Although these methods were illegal, 15% of fishers, mainly from Mijikenda origin, reported using them, but the true figure was probably higher (recently McClanahan, Maina & Davies, 2005b, reported a much higher figure).

Frequency of fishing differed greatly and included differences in duration of the fishing season and the number of weekly trips. Six times a week was mentioned most often. Fishing was divided into a high season of about 5.5 months and a low season of about four months, which left about two to three months when most fishers did not go out. The term 'high' season was ambiguous and this probably contributed to the large variation that was found. The high season was most commonly referred to as the season with the largest catches and this may differ for fishers, depending on their specialisation. The high season can also be defined as the season with the highest prices and, for popular species, this can be the time when catches were low, so that demand-supply interaction affected prices. The average number of annual trips was estimated at more than 300 per year.

Further analyses focused on four indicators, one for each of the selected features: willingness to stop fishing, inshore fishing grounds, destructive gear and annual number of trips. There was no significant relation between earner diversification and any of the fishing practices. Activity diversification correlated significantly with two selected indicators. Fishers with multiple activities used more destructive gear and fished the inshore grounds more often, while there was no sign that they were more willing to stop fishing for alternative employment. Apparently, activity diversification of fishers did not lessen the pressure on the marine environment. Rather, the opposite occurred in that fishers who had other employment onshore fished less prudently. Neither kind of diversification, apparently, provided fishers with the feeling that their fishing incomes had become any less important for survival. What emerged is that fishers with multiple activities fished on a smaller water area, used destructive gear more often and did not show more willingness to stop fishing for alternative employment.

7

Conclusions

Artisanal fishers on the Kenyan Coast face dwindling resources and heavy competition from tourism and human settlement, as is happening in many coastal areas in the third world. This will necessitate access to better fishing techniques and improved marketing facilities to continue with fishing as a means of livelihood and employment for local people. Sooner or later, however, fisher households, out of necessity, will have to broaden their resource base. Households that secure additional resources, notably non-maritime employment, strengthen their livelihood strategies in this way and improve household security.

The large majority of rural households in Kenya are smallholders who try to diversify with cultivation of food crops for home consumption, and income from cash crops, livestock and non-farm employment. Income diversification is an important strategy in household livelihoods and food security on the Kenyan coast as well (Hoorweg *et al.* 1995). Since fishers do not easily abandon the family profession, some form of resource diversification is the most likely strategy for them. As such, fishers who find economic alternatives and diversify their incomes will be less dependent on their catches. It was predicted that as a consequence, they will exert less pressure on marine resources and may even develop more positive attitudes towards conservation measures. Hopefully, they may even act as guardians and stockholders of the maritime heritage and become a positive force in environmental management.

Research into the social and economic conditions of fishers and their responses to the deteriorating situation is urgent. Firstly, this is so out of concern for the future of this group, which has thus far received little political attention. In all, there are an estimated 10-12,000 artisanal fishers on the Kenyan Coast, a relatively small number, although the total number of people who depend wholly or partly on fishing for their living may be as high as 190,000 if economic linkages and family dependents are taken into account. Secondly, out of environmental concern, since fishing activities can cause much damage to the reef and to the marine environment in general but also because fishers can be potential stewards of marine resources. Information and understanding of resource management and household strategies of local fishers are, therefore, vitally important.

Between 1999 and 2001, a team of researchers and students studied the relation between household livelihoods and resource conservation among artisanal fishers along the Malindi Coast, a stretch of about 125 km on the Indian Ocean. In all, the project was comprised of four main surveys and four support studies. The surveys covered the characteristics of fishers and fishing, fish landings, trading and marketing of fish, and livelihoods of fisher households. The support studies included catch composition and reproductive biology of fish, fish sales and marketing, income diversification of households and resource conservation by fishers respectively. The latter studies employed a mix of research methods including observation and catch records, in-depth interviews, participant observation, genealogies and life histories.

The research design allowed us to cover a broad area, not only in the geographical sense but also in the range of topics. The different surveys and studies also allowed independent confirmation of findings within the overall project. Because of the range of studies and the variety of methods that were used, the project also provided rich background information.

Limitations that have to be mentioned are that we relied, to a large extent, on verbal information from the fishers and their household members, which leaves the possibility of occasional misrepresentation. The surveys and studies were only loosely connected and did not follow a unified model that allowed straightforward aggregation or across the board comparisons. Sometimes, our conclusions were based on information from surveys with samples that were selected differently and that were not strictly comparable. Furthermore, the study attempted to relate concepts that were rather distant from each other: income diversification on the one hand and resource conservation on the other. These, at best, were indirectly related, which made it difficult to trace causal relations conclusively.

The main focus of the project was on income diversification of fishers, the pressure on marine resources and the relation between the two. In respect to income diversification, two issues, in particular, were given attention, namely how incomes of fishers compared with non-fishers and how diversification affected the incomes of fishers and the incidence of poverty among them. In respect to marine resources, attention focussed on the pressure on resources in terms of number of fishers, choice of fishing grounds, choice of gear and fishing intensity. And, finally, we asked whether there existed a relation between income diversification of fishers and pressure on marine resources, that is, whether fishers with income from more than one source ultimately exerted less pressure on the marine environment.

* * *

The marine fisheries in Kenya have the characteristics of many developing countries as listed by Hersoug, Jentoft & Degnbol (2004: 143), namely that (i) the management institutions are fragile; (ii) the fisheries are mainly non-industrialised with a widely spread population; and (iii) the resource base is a composite of many stocks that are harvested simultaneously. Most fishing was concentrated inshore. The large majority of fishing crafts consisted of different types of canoe and *dau*, which are suited mainly for reef and in-reef fishing. Vessels were either owned by one of the fishers or by a *tajiri* who leased the vessels to a captain usually against payment in kind. Modern gear, gill nets and lines were used most often while traditional gear such as traps and fences were on the decline. The gear was usually owned by the boat captain who was also responsible for maintenance of the equipment.

Two-thirds of the fishers visited other landing sites regularly because many moved along the coast in search of rich fishing grounds. Fishers took only some fish home but sold most of their catch. Two-thirds of the fishers sold the fish fresh but, at remote landing sites, the fish were often gutted and dried by the fishers themselves or by the fish traders. There were considerable local differences among fishers in the kinds of obstacles mentioned but lack of equipment was mentioned most often, followed by lack of funds and transport and marketing bottlenecks.

More than 100 fish species were identified at four selected landing sites. Although species richness was highest at the landing sites situated near Marine Protected Areas the amount of fish landed at these sites was considerably lower than at sites not near protected areas. Earnings per trip reflected the same differences, being highest at non-protected areas and lowest near protected areas. This confirmed earlier observations that Marine Protected Areas resulted in greater fish density but that more fishers were concentrated in smaller areas, resulting in lower catches (McClanahan & Mangi 2000).

Apart from catch differences among landing sites, there were also considerable seasonal differences. Catches were generally low between July and September but higher between October and April. Average earnings per crewmember per trip in the lowest quarter was only 60% of that in the highest quarter. Earnings from fishing showed large differences between seasons, among sites and among individuals.

Total income of fishers, which included earnings from fishing, farming and employment, was lowest in remote locations. In addition, fishers next to the Marine Protected Areas showed large differences in incomes from fishing. The lowest earnings per trip and lowest weekly incomes from fishing were reported by fishers at Uyombo. Earnings per trip at Mayungu were also low, but, weekly incomes from fishing were much higher at this site than at Uyombo. Income from agriculture and employment also varied, depending on agricultural potential of the area and employment opportunities, among others.

* * *

Fisher households were engaged in more economic activities than non-fisher households and they clearly had the advantage of a broader resource base. Fishers were also engaged in farming but had fewer farming assets than non-fishers. Income figures supported the above observations. Fisher households earned 50% more than non-fishers mainly due to earnings from fishing. Fishing was vital to the income of fisher households and, judging from the comparison with non-fishers, was difficult to compensate with other economic activities.

Theoretically, it is possible that the comparison group, the non-fishers living near the coastline, had income levels below rural communities elsewhere in Coast Province because of poor agro-ecological conditions. If that were the case, it is possible that fishers still had lower incomes than the population at large. However, this was shown not to be the case in that the incidence of poverty among fishers was not higher than among the general population in the rural parts of Coast Province. Therefore, the view that fishers are destitute, among the 'poorest of the poor', who are trapped in a hopeless situation and desperate for other employment opportunities did not apply here. Willman (2004) argued that poverty is not only decided by level of income but that fishers are often looked down upon because of low education, poor diets and low political participation. In the present study, fishers indeed had slightly less formal education than non-fishers, but there were only small differences in diets, differences that were not detrimental to fishers. Malleret-King (2003) found that fisher households near the Kisite Marine Park on the south coast were more food secure than others. This leads to the conclusion that the fishing sector did not function as an employer of last resort and that income diversification was not a means to escape from fishing but rather an integral part of livelihood strategies. The other side of the coin is that non-fishers will also be attracted to fishing in order to strengthen their livelihoods, and indeed, this was happening with the entry of a large number of Mijikenda fishers (Versleijen & Hoorweg 2008). Nevertheless, it must be kept in mind that many fisher households are still below the poverty line and, as such, are part of the general poverty problem in the region.

Two forms of income diversification have to be distinguished. Earner diversification refers to households with more than one income earner while activity diversification means that the head of the household has more than one economic activity. Activity diversification refers to an individual while earner diversification refers to a household. Few fisher households managed without diversification of some kind. Earner diversification resulted in substantially higher household incomes while activity diversification added considerably to the incomes of crewmembers but not in the case of boat captains. Earner and activity diversification can be regarded as different strategies, each with its own advantages and disadvantages. With activity diversification, the head can decide on the use of the income, but the income is limited by what one person can earn. It is reasonable to assume that possibilities for activity diversification decrease with age whereas possibilities for earner diversification increase with age. Wives can earn an income when older children are able to care for younger siblings. At a later stage, children may also earn money. Earner diversification means that the household has more income without the head putting in more work effort. The disadvantage from the head's point of view is that he does not have sole control of the income, although the wives and other income earners probably think differently. Still, it is likely that the wife will spend most or all her earnings on the household. As for older children, they, most likely, will marry and, therefore, contribute earnings only for a limited number of years.

Woodhouse (2002) postulated that diversification by an individual provides more flexibility in that it is likely to lessen risk. In the present study, fishers with activity diversification indeed showed adjustments to changing circumstances. However, diversification at the household level not only has the advantage of flexibility but also offers the possibility of specialisation for individual household members (Ellis 1999). In the case of earner diversification, fishers appeared to behave in this way indeed, not showing any signs of changing their specialisation.

There was not one clear-cut livelihood strategy that was best while the choices for or against diversification were complex. Fishers had to make a number of choices based on their personal circumstances and preferences. The first choice was whether to operate as a crewmember, an independent fisher or a boat captain. This choice depended on age, experience and personal initiative and had implications for the possibilities of income diversification. The second choice was whether to opt for activity diversification, which depended primarily on fisher status, job experience, level of education and personal preferences. For boat captains it was better to concentrate on fishing, but for crewmembers it was better to have alternative employment, as this group realized the highest incomes. Formal education may make it easier to find employment, although its importance should not be overestimated in rural circumstances, but chance opportunities were perhaps more deciding. A third choice was that of earner diversification, although this decision usually came later in life. The possibilities depended mainly on factors such as household phase (namely whether there are grown-up children present), household size and gender expectations. Above all, it should be realized that without income diversification of some kind, more fisher households would sink into poverty.

* * *

Overexploitation of natural resources is often associated with poverty among the local population, although state and commercial interests are often equally, if not more, responsible. Environmental degradation worsens the degree of poverty of marginal groups, which in turn leads to more intensive exploitation of accessible resources. The implicit assumption is that income improvements lessen the pressure on resources and may halt further damage to natural environments. Poverty itself has to be addressed; in particular, the poor have to be provided access to other sources of livelihood. However, the expectation that income improvements will halt environmental destruction has not generally been confirmed (Ellis 2000). People show great flexibility in finding and adding new opportunities to their repertoire without changing earlier practices.

Efforts to halt the downward spiral of poverty and resource degradation among fishers depend on the possibilities to improve the efficiency of small-scale fisheries, to enforce restricted access to some fishing grounds to conserve fish stocks, and to offer incentives to reduce fishing activities (Allison & Ellis 2001). State-imposed regulations to limit access have a high failure rate and tension exists between the two objectives of modern fishery policies, namely, increasing efficiency and regulating the catch. The weakness of implementing institutions in many developing countries also plays an important role (Alidina 2005).

The Department of Fisheries in Kenya has successively resorted under the Game Department, Min. of Tourism and Wildlife, Min. of Environment and

Natural Resources, Min. of Regional Development and the Min. of Livestock and Fisheries. In 2008, an independent Ministry of Fisheries has been instituted for the first time. A National Fisheries Policy has also been in preparation (Kenya 2007b). The key elements of the planned policies are: promoting co-management, stimulating public-private ownership, increasing domestic fish consumption, promoting fish trade, realizing local-value addition and striving after environmental integrity. Co-management serves to include fisheries managers, researchers, fishers, traders, processors and local communities in formulating management plans. The policy further recognizes the role of the private sector in fisheries development and the need to add more value to the fish products by local processing. The Kenya Government has also announced its intention to abolish the present licensing system for fishing in the EEZ and "replace it with Favourable Partnership Agreements with distant water fishing nations and fleets" with the aim "to develop the country's fisheries sector ... [as well as] ... conservation of the natural resources and ecosystem" (Kenya 2007b: 9).

Most fishers in the study area were aware of the degradation of marine resources and mentioned declining fish catches. They attributed this mainly to an increased number of fishers. This factor and three other elements of fishing activities that affect the marine environment were discussed, namely, the choice of fishing grounds, the choice of gear, and the intensity of fishing.

The number of fishers has been increasing over the past decades with the entry of the Mijikenda into the arena, a group not known for its fishing prowess until now (with the exception of the Digo on the south coast). Reasons for their entry included the open and easy access of the resource, the lax enforcement of license regulations and the need for employment. Half the fishers expressed an interest to opt for alternative employment although it is doubtful that they would abandon fishing completely if they found employment. It is more likely that they would try to combine the two, as was the case with many of the new entrants. Fishers with a low income were less willing to choose alternative employment, which is in line with the more general finding that families with higher incomes are usually in a better position to diversify than poorer families (Ellis 1999).

In developing countries, management of fisheries depends mainly on two sets of instruments (Allison & Ellis 2001), namely, controls to limit access (operating licenses, vessel capacity, closed seasons, closed zones) and measures to restrict efficiency or selectivity (gear prohibitions, mesh size regulation). On the Kenyan Coast, traditional restrictions on fishing grounds have fallen largely into abeyance (although they are still reported to exist on the south coast below Mombasa; see McClanahan *et al.* 1997). Their role has been taken by the Marine Protected Areas. The Marine Parks pose effective restrictions on fishing grounds and were mentioned often by fishers at nearby sites. However, they also had distinct disadvantages for the fishers, since they occupied good fishing grounds. Fishers showed considerable resentment against the authorities as a result. The need for conservation was not denied by fishers and most of them agreed that it was important for their future livelihoods. Many fishers said that they were prepared to support conservation projects but only if they would improve their living standards. Many fishers, especially those in Uyombo, found themselves in a situation that did not allow them to consider long-term consequences. Their main aim was to meet the short-term demands of their households.

The majority of fishers used nets with approved mesh sizes and hook and line (including long lines). Traditional gear have become less popular with time and were used by only one in ten of the fishers. Gear that are destructive include spear guns, beach seines and other nets with very small mesh sizes. Nets with small mesh sizes are particularly destructive as they catch non-targeted species and juveniles of the targeted species. Although these methods are illegal, 15% of fishers, mainly from Mijikenda origin, reported using them, but the true figure was probably higher (recently McClanahan *et al.* 2005b reported a much higher figure). Frequency of fishing differed and was affected by duration of the fishing season and the number of weekly trips. Six trips per week was the number most often mentioned. Fishing was divided into a high season of about 5.5 months and a low season of about 4 months, which left about two to three months without fishing (generally the height of the *kusi* season). The term 'high' season was ambiguous and this probably contributed to the large variation that was found. The average number of annual trips was estimated at more than 300 per year.

Ultimately, attention focused on four indicators, one for each of the selected features: willingness to stop fishing, fishing inshore grounds, using destructive gear and annual number of trips. If a relation between income diversification and fishing practices did exist, it should show primarily with activity diversification, since this affects the fishers more directly. An eventual relation between earner diversification and fishing practices would be expected to be less pronounced. There was no significant relation between earner diversification and any of the fishing practices. Activity diversification correlated significantly with two of the indicators. Fishers with multiple activities used more destructive gear and fished the inshore grounds more often, while there was no sign that they were more willing to stop fishing for alternative employment. Apparently, activity diversification of fishers did not lessen the pressure on the marine environment. Rather, the opposite occurred in that fishers who had other employment onshore fished less prudently. Neither kind of diversification, apparently, provided fishers with the feeling that their fishing incomes had become any less important to survive. What emerged was that fishers with multiple activities fished on a smaller water area, used destructive gear more often and did not show more willingness to stop fishing for alternative employment.

The results can be interpreted in several ways. It is likely that fishers who took up additional economic activities needed to stay inshore and were tempted to use non-legal gear in the hope of a quick catch. However, it is also possible that fishers who were active inshore and fishers who used non-legal gear tended to take up additional employment more often, although this is a less likely scenario. It can also be speculated that catches of inshore fishers were insufficient and forced fishers to find other work. Alternatively, fishers with only a fishing income needed to travel far out to sea to realise a sufficient income. But this did not account for 'new' fishers who neither had the equipment nor the experience needed for offshore fishing and who lacked knowledge of traditional practices. Surprisingly, activity diversification of fishers correlated with more destructive fishing practices, in contrast to expectations, which also lowers the positive environmental effects one may expect from policy initiatives aimed at generating employment opportunities.

* * *

We set out to explore the relation between household incomes and resource conservation. The first issue addressed was the income of fisher households and whether income diversification is of benefit to fishers. It was established that in this part of the Indian Ocean Coast fishers are better off than their immediate neighbours and are also not below the income level of the coastal population in general. Generally, income diversification was beneficial although this was not equivocal. Households with multiple earners had higher incomes than households with only a single earner. The impact of activity diversification depended on fisher status. Crewmembers with multiple activities managed to increase their incomes substantially although this was not so for boat captains and independent fishers.

The leading hypothesis was that income diversification would affect how fishers deal with marine resources and it was predicted that diversification would reduce the pressure on the marine environment. Fishers with alternative sources of income might be tempted to leave fishing or become less dependent on their catches. But this line of thinking was regretfully not confirmed. Apparently, there are no reasons inherent in the fishing sector for fishers to opt out and there is no reason to expect a decrease in number of fishers in the near future. Rather, indications are that the number of fishers is increasing with the entry of many (part-time) fishers. Still, it is possible that income diversification allows fishers to fish less intensively and be more prudent in the exploitation of the fisheries but this was also found not to be the case. In fact, fishers with alternative employment stayed inshore and used damaging gear more often, thus increasing the pressure on the delicate inshore environment, notably the coral reefs.

Attempts at income improvements among fishers in developing countries during the past decades have been based on the assumption that fishers are among the very poor (Payne 2000; Béné 2004). Improving catches and offering alternative employment are important elements of these approaches (Neiland 2004). The planned policy measures to combat existing poverty among fishers in Kenya include improvements of infrastructures such as landing sites, access roads, potable water supplies, fish sheds and electricity connections (Kenya 2001).[31] These measures are aimed primarily at strengthening the processing and marketing of fish, but there is no mention of creating employment opportunities for this group outside the fishing industry. Income diversification is certainly a means of improving the situation of fisher households – and not only among the fishers.

If employment opportunities were to be actively stimulated by government measures, there are two aspects that require careful consideration, namely, the type of employment and the geographical distribution. Employment opportunities within the fishing industry are limited by current catch levels that, already in 1996, were judged to be at maximum sustainable yields (McClanahan 1996). Employment opportunities outside the fishing industry will inevitably attract workers from outside the fishing communities as well. If the new industries are situated near Mombasa they will not offer easy access for fishers living a long distance away. But if the new industries are situated in more remote areas and near the coastline (to be in easy reach of the fishers), than it is likely that outsiders will follow. These newcomers will find accommodation locally and will realize the possibilities of taking up fishing for an extra income. We have learned that 'new' fishers usually stay near the coast, around the coral reefs. This will most certainly increase the pressure on this delicate part of the marine ecosystem, which is already under high pressure from tourism and pollution. A paradoxical scenario threatens in which employment opportunities designed to assist fishers will attract other people to the coastal strip where they will fish as an additional source of income and increase pressure on the marine environment.

Kenya is an atypical African country in that less than 10% of the national population lives within 100 km of the coastline, since the administrative and industrial centres of the country are situated inland. Worldwide, this percentage is close to 39% (Earthtrends 2008) and there is reason to expect that the coastal population of Kenya will increase rapidly because of influx from up-country provinces (Wakajummah 2000). The official projections are that the coastal population will increase by 18% between 2000 and 2010 and reach 3.2 million (Kenya 2002b).

It has been mentioned that employment generation is an urgent priority for Coast Province (Hoorweg *et al.* 2000). This is still very much the case; it would not only benefit fishers but also the far greater general population. Up to now, additional employment of fishers is mostly in agriculture with little wage employment. We do not know how many fishers would take up wage employment opportunities or how many newcomers would take up fishing but the interdependence of fishing with agriculture or employment should be recognized, as well as the need of a well integrated cross-sectoral development policy. It would be wishful thinking to expect that the pressure on marine resources will be lessened as a consequence of such developments. In fact, it is better to turn the argument around and to ask how to protect the marine environment from the growing pressure that will follow population growth and increased economic activities.

The coastal strip is characterised by ribbon-like habitation with an urban nexus in Mombasa and a few small towns along the coastline, like Kilifi, Malindi and Lamu. There is need for integrated approaches to fisheries, agriculture, water and other sectors that take into account large economic differences among landing sites, as well as seasonal variations in catches and large individual differences. Active employment policies will have to be designed carefully in terms of industry location, labour requirements and environmental pollution that take into account the protection of the coastal shores and reefs. This requires a degree of fine-tuning that may be unrealistic and beyond the power of existing government agencies.

For the protection of the marine environment from intensive fishing, solutions will have to be sought in other directions such as access restrictions and countering destructive practices. Marine Protected Areas are a proven way to restrict access to certain areas but implementing agencies should be aware of the large amount of resentment they generate among the fishers nearby. The general feeling among fishers is that they have been deprived of their best fishing grounds for which they have received little compensation. Apart from the total ban in the Parks, it is recommended that access also be restricted in the Reserves where fishing is still allowed. The number of fishers in these areas is theoretically controlled through government fishing licenses and the approval of fisher committees. Neither method of restriction works well at the moment as the enforcement of government regulations is lax and the local fishers are not keen to deny others access to an income except in exceptional circumstances. Responsibility should be in the hands of the Fisheries Department and the focus should be on restricting the large number of new fishers that enter the arena by toughening license requirements and inspections.

Difficult as it may be to regulate the number of fishers, it is even more difficult to control fishing practices. Collaboration with fishers and local population is needed not only for effective resource management but also to incorporate traditional environmental knowledge in the management of fisheries (Mathooko 2005). Effective management can only be shaped at the community level and successful management strategies of fishery resources must be adopted by the communities concerned. The fisheries sector has, by now, considerable experience with co-management as institutional arrangement to share management responsibility among stakeholders (Wilson, Nielsen & Degnbol 2003; Pomeroy & Rivera-Guieb 2006). Co-management goes beyond community-based management because it involves decentralisation and sharing of responsibilities between local government and user groups to different degrees (Pomeroy & Berkes 1997). Resource users, the fishers, have to be involved in decision-making and drafting regulations and it is important that they have a positive attitude towards conservation (Gelchich, Edwards-Jones & Kaiser 2005). Fishers generally do not condone damaging practices since they are well aware of the dangers to their livelihoods in the long-term. The need for conservation measures was not denied by the fishers and most of them agreed that it was important for their future livelihoods. But many fishers were in a situation where they could not afford to consider the longer term but had to meet the short-term demands of their families. Many fishers expressed a willingness to participate in marine conservation projects but they would only do so if they could expect income improvements in the short-term and have confidence in the long-term prospects. Ultimately, this implies a shift of responsibility from government agencies towards local communities. The fishers will have to be involved in setting the agenda and attention must be given to the priority areas that fishers themselves consider important.

Appendices

Appendix 1 List of fish species in the Malindi-Kilifi marine waters

English/Common name	Kiswahili/Local name	Latin name
Anchovy	Dagaa/Mcheli	*Anchovieli indica*
Angelfish	Kitatange	*Honiochus acuminatus*
Baraccuda	Tengezi	*Sphyraena japonica*
Baraccuda	Tengezi	*Sphyraena jello*
Batfish	Tuguu	*Platax Pinnatus*
Blackskin	Fute	*Gaterin sordidus*
Bonito/Skipjack	Jodari	*Euthynnus pelamis*
Butterflyfish, threadfin	Kikorokoro	*Chaetodon auriga*
Caesio	Viunda/Mweru	*Caesio xanthonotus*
Catfish, eel	Ngogo/Mtonzi	*Plotosus arab*
Catfish, striped eel	Ngogo/Mtonzi	*Plotosus lineatus*
Cavillajack	Kisukari	*Elagatis bipinnulata*
Chubfish, brassy	Kufi/Kimbulimbuli/Kukusi	*Kyphosus vaigiensis*
Coris, queen	Mwenza mawe	*Coris formosa*
Damselfish, black	Patima mashowera	*Stegastes nigricans*
Damselfish, false-eye	Patima mashowera	*Abudefduf sparoides*
Emperor, blackspot	Mchakufa	*Lethrinus harak*
Emperor, spangled	Changu macho	*Lethrinus nebulosus*
Emperor, variegated	Changu	*Lethrinus variegatus*
Filefish, barred	Gona/Sharifu	*Cantherhines dumerilli*
Filefish, broom	Gona/Sharifu	*Amanses scopas*
Filefish, honeycomb	Puju	*Cantherhines pardalis*
Filefish, spectacled	Gona/Sharifu	*Cantherhines fronticinctus*
Flathead fish	Vumbama	*Platycephalus crocodila*
Goatfish, dash-dot	Mkundaji	*Parupeneus barberinus*
Goatfish, double bar	Mkundaji	*Parupeneus bifasciatus*
Goatfish, Indian	Mkundaji	*Parupeneus indicus*
Goatfish, yellow stripe	Sonyo	*Mulloides flavolineatus*
Grey skin	Fute moshi	*Gaterin batata*
Grouper, jewel	Tewa ndudu	*Cephalopholis miniata*
Grouper, peacock	Tewa shambaru	*Cephalopholis argus*
Grouper, redbanded	Tewa	*Caeruleo punctatus*

Appendix 1, continued

English/Common name	Kiswahili/Local name	Latin name
Grouper, redbanded	Tewa	*Epinephelus fasciatus*
Grouper, squaretail	Tewa moshi	*Plectropomus areolatus*
Grunter, spotted	Tamamba	*Pomadasys operculare*
Halfbeak	Chuchungi/Kidau	*Hemiramphus far*
Kingfish	Nguru	*Scomberomorus commerson*
Kingfish, blacktip	Kambisi	*Caranx sem*
Lemonfish	Nyeya	*Gaterin gaterinus*
Mackerel, little	Oona	*Rastrelliger kanagurta*
Marlin, black	Sulisuli	*Makaira indica*
Marlin, blue	Sulisuli mviringo	*Makaira nigricans*
Milkfish	Mwatiko	*Chanos chanos*
Minstrel	Fute	*Plectorhinchus schotaf*
Moony, silver	Pakawe	*Monodactylus argenteus*
Mullet	Mkizi	*Mugil cephalus*
Needlefish, crocodile	Mtumbuu	*Tylosurus crocodilus crocodilus*
Needlefish, yellow	Mtumbuu	*Strongylura leiura*
Parrotfish	Pono mwamba	*Callyodon guttatus*
Parrotfish, bullethead	Pono	*Scarus sordidus*
Parrotfish, christmas	Pono kasiki	*Calotomus carolinus*
Parrotfish, marbled	Pono	*Leptoscarus vaigiensis*
Parrotfish, stareye	Pono	*Calotomus carolinus*
Pursemouth	Chaa	*Gerres oyena*
Rabbitfish, forktail	Tafi mtunga	*Siganus argenteus*
Rabbitfish, starspotted	Tafi manga	*Siganus stellatus*
Rabbitfish, whitespotted	Tafi	*Siganus sutor*
Ray, manta	Taa chui	*Manta birostris*
Ribbonfish	Panga	*Trichiurus lepturus*
Rock cod	Chewa/Tewa	*Epinephelus merra*
Rubber lip, blackspotted	Mleya/Nyeya	*Plectorhinchus gaterinus*
Runnerfish	Songoro	*Rachycentron canadus*

Appendix 1, continued

English/Common name	Kiswahili/Local name	Latin name
Sailfish	Sulisuli	*Istiophorus platypterus*
Sailfish	Sulisuli makuti	*Istiophorus gladius*
Sardine	Simu	*Sardinella melaneura*
Sawfish, largetooth	Papa upanga	*Pristis microdon*
Scavenger	Nyavi	*Lethrinus miniatus*
Sergeant fish, scissortail	Patima mashowera	*Abudefduf sexfasciatus*
Shark, basking	Papa usingizi	*Rhincodon typus*
Shark, blacktip reef	Papa	*Carcharhinus melanopterus*
Shark, tiger	Zambarani	*Galeocerdo cuvier*
Sicklefish, concertina	Shana	*Drepane longimanus*
Snapper, black spot	Tembo/Kungu	*Lutjanus ehrenbergii*
Snapper, blood	Tembo/Kungu	*Lutjanus sanguineus*
Snapper, blotcheye	Kifuvu/Kibaazi	*Myripristis murdjan*
Snapper, blue banded	Tembo-uzi	*Lutjanus kasmira*
Snapper, dory	Tembo/Kungu	*Lutjanus fulviflamma*
Snapper, hump-back	Runga/Ndawasho	*Lutjanus gibbus*
Snapper, one spot	Tembo/Kungu	*Lutjanus monostigma*
Snapper, speckled	Cheusi	*Lutjanus rivulatus*
Snapper, two-spot red	Tembo/Kungu	*Lutjanus bohar*
Soldierfish	Kibaazi/Kifu	*Holocentrus summara*
Spadefish	Tuguu/Kudusi	*Platax orbicularis*
Stingray, blackspotted ribbontail	Nyenga	*Taeniura melanospilos*
Stingray, bluespotted ribbontail	Nyenga	*Taeniura lymma*
Streaker	Mshikashangwi	*Aprion vireucens*
Surgeonfish, convict	Kangaja	*Acanthurus triostegus*
Surgeonfish, powder-blue	Kangaja	*Acanthurus leucosternon*
Sweeper	Makarenge	*Pempheris ovalensis*
Sweeper, black-edged	Makarenge	*Pempheris mangula*
Sweetlips, black spotted	Mchone/Mleya/Kumba-maji	*Plectorhinchus gaterinus*
Sweetlips, grey	Mleya	*Plectorhinchus schotaf*
Tripletail	Kanda/Stefua	*Lobotes surinamensis*

Appendix 1, continued

English/Common name	Kiswahili/Local name	Latin name
Thornfish, straight-lined	Ngagu	*Terapon theraps*
Trevally, bluefin	Kolekole	*Caranx melampygus*
Trevally, bluefin	Kolekole	*Caranx stellatus*
Triggerfish, half moon	Kikande	*Sufflamen chrysopterus*
Triggerfish, red tooth	Kikande	*Odonus niger*
Trumpetfish	Mzu-moshi	*Aulostomus chinensis*
Tuna, yellowfin	Jodari	*Thunnus albacares*
Unicornfish, spotted	Puju	*Naso brevirostris*
Wahoo	Nguru ngazija	*Acanthocybium solandri*
Wrasse, cigar	Mbooya mvuvi	*Cheilio inermis*
Wrasse, goldbar	Bua	*Thallassoma hebraicum*
Wrasse, tripletail	Stefua	*Cheilinus trilobatus*
CRUSTACEA		
Crab	Kaa	*Brachyura*
Lobster, ornate spiny	Kamba mawe	*Panulirus ornatus*
Lobster, painted spiny	Kamba mawe	*Panulirus versicolor*
Prawns	Kamba wadogo	*Penaeus indicus*
MISCELLANEOUS		
Bêche-de-mer	Jongoo la pwani	Various
Octopus, whitespotted	Pweza	*Octopus macropus*
Squid, big-fin reef	Ngisi nyamvi	*Sepioteuthis lessoniana*
Squid, Indian	Ngisi	*Loligo duvaucelli*

Source: Catch Survey, Mohammed (2002) and Glaesel (1997b).

Appendix 2 Catch composition by landing site (%)
(Number of times species were present in catch records; 0.5% or more)

English	LS Ngomeni (N=1227)*	LS Mayungu (N=1516)	LS Uyombo (N=1751)	LS Takaungu (N=942)	Total (N=5436)
Rabbitfish	3.4	59.0	75.8	17.4	44.6
Emperor	2.1	19.3	64.7	21.8	30.5
Parrotfish	0.4	1.6	32.6	3.2	11.6
Snapper	2.9	1.6	23.4	16.8	11.5
Goatfish	-	0.1	33.1	0.4	10.8
Wrasse	-	0.2	28.7	0.2	9.3
Mullet	35.5	0.1	0.7	3.9	8.9
Rock cod	2.8	1.6	15.2	4.4	6.7
Kingfish	12.8	3.1	0.7	10.8	5.8
Ribbonfish	0.1	0.1	0.1	30.2	5.3
Sweetlips	-	-	14.5	-	4.7
Shark	12.5	2.4	0.3	5.3	4.5
Squid	-	9.7	2.9	1.3	3.8
Lobster	11.8	0.5	0.1	0.4	2.9
Pursemouth	-	0.3	5.6	0.5	2.0
Barracuda	0.2	0.2	1.5	6.4	1.7
Crabs	6.7	-	0.1	-	1.5
Surgeonfish	0.1	-	4.5	-	1.5
Octopus	-	2.4	1.2	2.2	1.4
Scavenger	1.8	3.5	-	-	1.4
Minstrel	-	0.2	2.8	1.3	1.2
Halfbeak	0.2	0.1	2.9	-	1.0
Rayfish	-	-	1.4	2.7	0.9
Damselfish	-	-	2.7	-	0.9
Spotted Grunter	0.3	0.2	1.6	0.7	0.8
Catfish	1.3	0.1	1.0	0.2	0.7
Tuna	-	2.2	0.1	-	0.6
Triple tail wrasse	-	-	1.7	-	0.6
Soldierfish	-	-	1.7	-	0.6
Butterflyfish	0.1	0.1	1.3	0.3	0.5
Prawns	2.2	-	-	-	0.5
Unicornfish	-	0.7	1.0	-	0.5
Total number of species recorded	28	35	47	31	63

Source: Catch Survey.
* N = Number of weighted catch records

Appendix 3 Household food consumption

Introduction

Cereals constitute the main staple food in Kenya and this is also the case in the coastal region. In general, maize is most important while traditional cereals such as millet and sorghum are on the decline. At the coast, cassava and rice are also popular staple foods. In most households, three meals, which consist of freshly prepared foods and/or leftovers, are served daily. Some households skip breakfast and/or lunch in times of food shortage. Most household members return home for dinner. Maize is usually eaten as stiff porridge (*ugali*) together with a relish that may consist of legumes, green vegetables, tomatoes, onions and/or products of animal origin. *Ugali* is served on a large plate or bowl and is eaten by breaking off lumps and dipping it in the relish. In most parts of the country, the food is rather monotonous but coastal food dishes have a greater variety in that they contain more ingredients – notably fish, coconut and spices. Detailed information on food consumption in six coastal areas was compiled earlier during a comprehensive research project that was reported in detail (Niemeijer, Foeken & Klaver 1994; Hoorweg *et al.* 1995; Klaver & Mwadime 2000)

Method

Food consumption is usually studied by recall or observation methods. Household observations are laborious in data collection as well as data analysis. Recall methods, such as the 24-hr recall, may be less laborious in terms of data collection but analysis is equally taxing. Both methods have restrictions in the time period covered (usually 1 or 2 days) and the scope of information (restricted to the food that is eaten in the household under observation, not what is eaten elsewhere).

Food consumption data were collected for the 213 households visited in the household survey. Since the primary focus of the household survey was on economic activities and fishing practices, only limited attention could be given to food consumption. Data collection was curtailed and respondents were asked questions about three food categories:

- staple foods, legumes/vegetables and animal products.

For each food category, respondents were asked to mention:

- the three most common foods,
- how often these foods were prepared during the week and
- whether the food was home cultivated or purchased.

Respondents were also asked whether

- any staple foods were stocked and, if yes,
- the number of months these staple foods were expected to last.

Table A3.1 Food practices among group of fishers and non-fishers (N=213)

		Consumption (%) 1	Frequency (average) 2	Household production (%) 3	Amount of food stock (average) 4	Hhlds with food stock (%) 5
Staple foods	Maize	99.1	6.19	60.1	2.25	60.1
	Cassava	66.7	1.81	39.1	2.30	38.0
	Rice	67.1	1.53	1.4	0.0	–
	Wheat	23.9	0.69	0.0	0.0	–
	Other	1.9	–	–	–	–
Legumes & vegetables	Beans	76.1	2.39	2.3	–	–
	Peas	54.5	2.08	43.2	–	–
	Greens	90.2	3.17	53.1	–	–
	Other	5.2	–	–	–	–
Animal products	Fish	99.1	5.80	64.8	–	–
	Beef	52.1	0.73	1.4	–	–
	Chicken	49.8	0.51	41.3	–	–
	Eggs	7.0	0.12	3.8	–	–
	Other	–	–	–	–	–

Source: Household Survey
1. H'hlds mentioning foodstuff as one of three most common foods in respective food groups
2. Number of times that food is reportedly eaten during the week (average)
3. Percent households reporting home production of foodstuff (%)
4. Estimated duration of stock of staple food (average in months)
5. Households reporting stock of staple food (%)

Results

Staple Foods. Three staple foods dominated: maize, cassava and rice. All households consumed maize in the course of the week, while cassava and rice were consumed by two-thirds of the households (Table A3.1).

There were up to 10 staple food dishes prepared a week: maize was highest with 6.1 times, followed by cassava and rice (between 1-2 times each) and wheat flour (less than once a week; used to prepare *chapati*).

Maize was cultivated by about 60% of the households and cassava by about 40%. Very few households cultivated rice and virtually all the rice and wheat flour were purchased.

About 60% of the households had stock of maize but only about 30% had stock of cassava.[32] The stock for the total group would last an average of 4.5 months, with equal contributions from maize and cassava.[33]

Legumes/Vegetables. Beans and peas, the two main legumes, were prepared by three quarters and one half of the households, respectively. In total, legume dishes were consumed an average of 4.4 times/week; 2.4 and 2.1 times per week for beans and peas, respectively. Half the households cultivated peas, but beans were nearly always purchased. The coastal climate is more favourable to growing peas than beans.

Greens were prepared by almost 90% of the households during the week with an average number of 3.1 times/week. About 54% of the households grew green vegetables while about 47% bought them.

Animal Products. The main animal products were fish, beef and chicken. Fish was consumed most frequently and was prepared in nearly all households; half the households consumed beef and chicken. Fish dishes were eaten 5.8, beef 0.7 and chicken 0.5 times/week while eggs were eaten even less frequently. For the group as a whole, fish was most often from own catch (64.8%); all fishers took fish home while non-fishers bought this food. Beef was usually bought whereas chicken was most often from the household's own flock.

Food security indicators

Food security entails a number of components, including food production, food stocks, food consumption and food composition. Using the information described above, three types of indicators were constructed to assess the food security situation in the households.

The first indicator assessed the home production of foods. It was defined as the percentage of households that produce one or more of the foods that were mentioned by respondents as commonly eaten for each of the respective food groups (staple foods, legumes/greens, animal products). In the case of staple foods, 60% of the respondents reported that they produced one or more of the staple foods. The percentage was similar (59%) for legumes and greens. Animal products were considered to be home produced when they were from domestic animals being kept by the household or were caught, as in the case of fish. This figure was 80%, mainly a consequence of the frequent consumption of fish in this particular population.

The second indicator assessed food stocks, notably whether there was any stock of staple food in the household, the estimated amount of stock or how many months the stock was expected to last. More than half the households, 61%, reported that they had some food stock consisting of maize and/or cassava. The food stock was expected to last an average of 4.4 months for the population as a whole, but 7.3 months for only the households that actually had stock.

The third type of indicator assessed the food consumption, namely the number of times that foods were prepared during the week. Weekly, staple food dishes were prepared about 10 times, legumes/greens dishes 7.6 times and animal products 7.2 times. From observations and other studies, the amount of animal products consumed was small.

Conclusion

Table A3.2 presents the aggregated scores of the three indicators of the sample population for the respective food groups. In Chapter 5, these indicators were used to examine whether food security differs with fisher status and/or with income diversification.

Table A3.2 Food security indicators among group of fishers and non-fishers (N=213)

Food production: Percent households with home production of one or more of the foods mentioned	
Staple foods (%)	60.1
Legumes & greens (%)	58.7
Animal products (%)	80.3
Food stock: Percent households reporting stock of staple foods & estimated duration of stock (months)	
Households with food stock (%)	60.6
Size of food stock (ave.) [3]	4.4
Food consumption: Average number of times that foodstuffs are prepared as dish in course of one week	
Staple foods (ave.)	10.2
Legumes & greens (ave.)	7.6
Animal products (ave.)	7.2

Source: Household Survey

Appendix 4 Regulations in Marine Parks and Marine Reserves

Within the Marine National Parks and Reserves certain regulations are in force. There regulations are not only concerned with conservation but also with the access and the use of the area. For example, it is not allowed to:

- Engage in any of the following marine activities without paying the prescribed fees:
  - Goggling
  - Water skiing
  - Diving
  - Site viewing in Mida Creek
  - Operate or use a glass-bottom boat or any other marine vessel in the Park area;
- Reside in the Marine National Park and Reserve;
- Clear any vegetation in the Marine National Park and Reserve;
- Posses any weapons, explosives or poison in the Marine National Park;
- Collect shells, aquarium fish and corals in Marine National Reserve;
- Kill or capture any mammal or turtle;
- Harass any mammal or turtle so as to disturb its behaviour or breeding grounds;
- Chase any marine mammal or turtle with intent to kill;
- Take any marine mammal or turtle, alive or dead, including any marine mammal or turtle stranded on land;
- Remove any marine animal or vegetation or alter existing forms of prehistoric, archaeological, historical or other scientific interest in the Park area;
- Use the following prohibited methods while fishing in the Reserves:
  - Trawling within the Marine Reserve
  - Use of spears for fishing
  - Use of any explosives, poisonous or noxious substances or electric shocks for the purpose of rendering fish more easy to catch;
- Fish in the Marine Park.

An exception to these regulations may be obtained through a special permit from the director of the KWS or an Officer authorised by him.

Source: Hof 1999; Versleijen 2001

Notes

1 Major Ewart Grogan, famous for his Cape to Cairo foot journey and other colonial ventures (Paice 2001).
2 The essential elements of small-scale fisheries are generally defined as "labour intensive harvesting, processing and distribution ... conducted full-time, part-time or seasonally ... to supply fish and fishery products to local and domestic markets and for subsistence consumption" (FAO 2004b). These elements are little different from the definition of artisanal fisheries given in the main text.
3 This estimate is based on a multiplier of 2.5 for workers in support industries and an average (rural) household size of 4.76 (Kenya 2002a: 38).
4 The use of poison was already mentioned as a widespread problem along the coast by the Assistant Fish Warden in 1949 (Kenya 1950b).
5 Diani Marine Reserve was officially gazetted in 1995, but the authorities are treading carefully in view of the resistance among the local population and the time of writing there was no management plan in place (McClanahan *et al.* 2005a).
6 The Kilifi coastal tract was part of the Fisher and Trader Survey but was not included in the other research activities.
7 Additional details on survey methods and sample characteristics can be found in Hoorweg, Degen & Wangila (2003).
8 Certain questions allowed for more than one answer by the respondent. This is indicated in the tables concerned with MR (Multiple Response); in these cases totals may add up to more than 100%.
9 Lists were made of fishers who appeared more than 20 times (20+), 15 times (15+) and 10 times (10+) in the catch records until then. Selection concentrated at first on fishers (20+); once this group was exhausted fishers (15+) were accepted and finally fishers (10+).

10 Boat captains (N=57) enlisted crewmembers for their trips. Independent fishers (N=26) operated alone, they mostly had small boats; a few went on foot or swam.

11 This figure is larger than the 3.0 persons per boat reported 20 years earlier by Okidi (1979: 15-16).

12 The locally used word *dau* or *dhau* refers to specific vessels; the word *dhow* common in English usage refers to a much broader class of vessels including *dau*, *mashua* and *jahazi*. Therefore, a *dhow* is not always a *dau*.

13 The tables do not include figures for shellfish although they were recorded when the assistants came across them. The average price paid for octopus was Ksh 56 (N=46), for crab it was Ksh 130 (N=17), for prawns Ksh 234 (N=26) and lobster Ksh 370 (N=34). The selling margins were uniform at 32% except for octopus, which was rather high at 62%.

14 The poverty line in this case was Ksh 1239 per person per month in the rural areas and Ksh 2648/month in the urban areas (Kenya 2001: 13). For the rural areas this amounts to US$ 19.8/month or US$ 0.66/day. (The exchange rate in 1997 was US$ 1.0=Ksh 62.7; Kenya 2003).

15 Recent estimates show that Coast Province remains among the provinces with a high poverty incidence of 58% (Kenya 2005).

16 In the other fisher households, it was one of the other household members who was engaged in fishing (see Chapter 2).

17 Fish trading, perhaps surprisingly, was not an important activity in the households surveyed except among the crewmembers where 15% of the wives were engaged in fish trading or processing.

18 There was only one household where this was not a grown-up child.

19 In the text, the terms 'employment' and 'employment activities' refer to self-employment and wage employment together.

20 Analysis of variance (N=187; dependent variable: household income; covariate: household size):

	df	F	p
Covariates	1	7.19	.01
Main effect	1	15.27	.00

21 The incomes calculated in this study did not include the value of subsistence crops grown for own consumption. If these had been included, the comparison with the general population would have been even more favourable.

22 In statistical terms: 48% of the cases had the same score on activity diversification and earner diversification (yes/yes or no/no) and 52% of the cases had a different score (yes/no and no/yes).

23 Analysis of variance (N=111; dependent variable: household income; covariate: household size):

	df	F	p
Covariate	1	2.47	.12
Main effects	2	2.31	.10
Activity diversification	1	0.01	.91
Earner diversification	1	4.59	.03
2-way interactions	1	0.15	.70

24 Reasons for the lower official estimate of the number of fishers in Kilifi and Malindi were that only official landing sites were covered and that many fishers did not possess a fishing license.

25 None of the fishers admitted using poison but it was learned reliably that it was used in the far northern parts of the Malindi District, near Mto Kilifi.

26 Again, none of the fishers admitted to this, but dynamite was reportedly used sometimes between Mayungu and Watamu.

27 Fisher Survey: Mijikenda (25.0%) vs. Bajun (2.4%).

28 Some fishers preferred to set out during the weekends because the frequency of government patrols was reportedly lower at this time.

29 An interesting phenomenon that was noted in Mayungu was that of 'joy' fishing (analogy of 'joy' riding). Since most fishers at this landing site lived inland, their boats were left largely unattended and other fishers sometimes used the boats during the night without permission of the owners.

30 There was also a significant difference in the number of landing sites frequented; the fishers with 'multiple' activities reporting fewer landing sites, understandable if they have other work to do on land (ANOVA, df=1, F=14.0, p=.00).

31 This refers to the measures mentioned in the Poverty Reduction Strategy Paper 2001-2004, which was, however, never implemented because of a stand-off between government and donors (Freeman, Ellis & Allison 2004).

32 Since rice and wheat flour were purchased, there were no stocks recorded.

33 When the stocks of maize and cassava were combined, it was found that 30-40% of the households did not have any stock, 30% had stock of less than 6 months and 32% of the households had combined stock of 6 months or more.

References

ABDULLAH, A., S. ALI HAMAD, A. MBAROUK ALI, & R.G. WILD (2000). Misali Island, Tanzania: An open access resource redefined. Paper presented at *8th Biennial IASCP Conference; Constituting the commons: Crafting sustainable commons in the new millennium.* Bloomington: International Association for the Study of Common Property. (htpp//dlc.dlib.indiana.edu/archive/00000193).

ACHESON, J.M. (1981). Anthropology of fishing. *Annual Review of Anthropology* 10: 275-316.

ALIDINA, H.M. (2005). Local level fisheries management in Diani-Chale, Kenya: Current status and future directions. *Coastal Management* 33 459-470.

ALLISON E.H. & F. ELLIS (2001). The livelihoods approach and management of small-scale fisheries. *Marine Policy* 25: 377-388.

ALOO, P. (2000). Marine resources. In J. Hoorweg, D. Foeken & R.A. Obudho, eds., *Kenya Coast Handbook: Culture, resources and development in the East African littoral,* pp. 43-60. Hamburg: LIT Verlag.

— ed. (2007). *Fishery industry in Kenya: Towards the development of a national policy.* Nairobi: Ministry of Livestock and Fisheries Development/Food and Agriculture Organization.

ARDILL, J.D. & M.J. SANDERS, eds. (1991). *Proceedings of the seminar to identify priorities for fisheries management and development in the southwest Indian Ocean, Albio, Mauritius, 3-5 Sept. 1991.* (Fisheries Report No. 457). Rome: Food and Agriculture Organization.

BÉNÉ, C. (2003). When fishery rhymes with poverty: A first step beyond the old paradigm on poverty in small-scale fisheries. *World Development* 31 (6): 949-975.

— (2004). Poverty in small-scale fisheries: A review and some further thoughts. In A.E. Neiland & C. Béné (eds). *Poverty and small-scale fisheries in West Africa*, pp. 59-102. Dordrecht: Kluwer Academic Publishers.

BÉNÉ, C., G. MACFADYEN & E.H. ALLISON (2007). *Increasing the contribution of small-scale fisheries to poverty alleviation and food security*. (FAO Fisheries Technical Paper 481). Rome: Food and Agriculture Organization.

BULTE, E. (1997). *Essays in economics of renewable resources*. Wageningen: Wageningen Agricultural University.

CDA (1996). *Towards integrated management and sustainable development of Kenya's coast*. Mombasa: Coast Development Authority.

— (2001). *Moving coastal management forward: Kenya progress report 1994-1999*. Mombasa: Coast Development Authority.

CHAMBERS, R. (1987). *Sustainable livelihoods, environment and development: Putting poor rural people first*. (IDS Discussion Paper 240). Brighton: Institute of Development Studies.

CHARLES, A. (2001). *Sustainable fishery systems*. (Fish and aquatic resources series 5). Oxford: Blackwell.

CHAYANOV, A.V. (1966). The theory of peasant economy. In D. Thorner, B. Kerblay & R.E.F. Smith, eds., *The theory of peasant economy*, pp. 29-269. Homewood, Ill.: Richard Irwin.

CHILUNDO, A.G. & B.M. CAU (2000). Traditional forms of common property rights: a case study in Southern Mozambique. Paper presented at *8th Biennial IASCP Conference; Constituting the commons: Crafting sustainable commons in the New Millennium*. Bloomington: International Association for the Study of Common Property. (htpp//dlc.dlib.indiana.edu/archive/00000235).

CINNER, J.E. & T.R. MCCLANAHAN (2006). *A baseline socio-economic assessment of fishing communities along the north coast of Kenya*. Mombasa: Wildlife Conservation Society's Coral Reef Conservation Project.

DAILY NATION (1998). Fishing methods queried. *Daily Nation*, October 7.

— (2007). Fishermen cast nets deeper and wide, for less. *Daily Nation*, December 6.

DEPARTMENT OF FISHERIES (1996-1998). Fisheries records 1996, 1997, 1998.

DUBBINK, W. & M. VAN VLIET (1996). Market regulation versus co-management? Two perspectives on regulating fisheries compared. *Marine Policy* 20 (6): 499-516.

EARTHTRENDS (2008). Coastal and marine ecosystems > Fisheries – population within 100 km of coast. (earthtrends.wri.org; accessed on 21.10.1008).

EAST AFRICAN (2000). Tanzanian fishermen dynamiting fish in Kenya's waters. *East African*, October 23.

ELLIS, F. (1998). Household strategies and rural livelihood diversification. *Journal of Development Studies* 35 (1): 1-38.

— (1999). *Rural livelihood diversity in developing countries: Evidence and policy implications*. (Natural Resource Perspectives 40). London: Overseas Development Institute.

— (2000). *Rural livelihoods and diversity in developing countries*. Oxford: Oxford University Press.

ELLIS, F. & H.A. FREEMAN (2004). Rural livelihoods and poverty reduction in four African countries. *Journal of Development Studies* 40 (4): 1-30.

FAO (1966a). *The fisheries of East Africa and their development prospects: A preliminary survey*. (Fisheries Circular 103; Prepared by A.A. Kerr). Rome: Food and Agriculture Organization.

— (1966b). *Report to the Government of Kenya of a survey on longline fishing resources in East African waters*. (Report FAO/UNDP 2191). Rome: Food and Agriculture Organization.

— (1969). *Report to the Government of Kenya on investigations into pelagic, demersal and crustacean resources of the coast of Kenya.* (Report FAO/UNDP 2733). Rome: Food and Agriculture Organization.

— (1971). *Report to the Government of Kenya on the evaluation of prawns (Penaeidae) and spiny lobster (Palinuridae) marine resources.* (Report FAO/UNDP 3006). Rome: Food and Agriculture Organization.

— (1979). *Report of the FAO/IOP workshop on the fishery resources of the western Indian Ocean south of the equator. Mahe, Seychelles, Oct 23-Nov 4.* (Indian Ocean Programme, Development Reports 45). Rome: Food and Agriculture Organization.

— (1983). *Kenya, offshore trawling survey project findings and recommendations.* Report prepared for the Government of Kenya. Terminal Report. FI:DP/KEN/74/023. Rome: Food and Agriculture Organization.

— (1984). *Fisheries report.* (No. 320). Rome: Food and Agricultural Organization.

— (1985). FAO Species identification sheets for fishery purposes. Western Indian Ocean: Fishing Area 51,Vol. IV. Rome: Food and Agricultural Organization.

— (1996). *The state of the world fisheries and aquaculture 1996.* Rome: Food and Agriculture Organization

— (2002). *The state of the world fisheries and aquaculture 2002.* Rome: Food and Agriculture Organization.

— (2004a). *The state of the world fisheries and aquaculture 2004.* Rome: Food and Agriculture Organization.

— (2004b). *Report of the second session of the working-party on small-scale fisheries, Bangkok, 18-21 November 2003.* (FAO Fisheries Report No. 735). Rome: Food and Agriculture Organization. (Cited in Béné *et al.* 2007: 8-9)

— (2007). *The state of the world fisheries and aquaculture 2006.* Rome: Food and Agriculture Organization.

FOEKEN, D. (2000). Physical resources and infrastructure. In J. Hoorweg, D. Foeken & R.A. Obudho, eds., *Kenya Coast Handbook: Culture, resources and development in the East African littoral,* pp. 29-42. Hamburg: LIT Verlag.

FRAZIER, J.G. (1993). Dry coastal ecosystems of Kenya and Tanzania. In E. v.d. Maarel, ed., *Ecosystems of the world. Vol. 2B: Africa, America, Asia and Oceania*, pp. 129-150. Amsterdam: Elsevier.

FREEMAN, H.A., F. ELLIS & E. ALLISON (2004). Livelihoods and poverty reduction in Kenya. *Development Policy Review* 22 (2): 147-171.

FULANDA, B. (2003). Shrimp trawling in Ungwana Bay: A threat to fishery resources. In J. Hoorweg & N. Muthiga, eds., *Recent advances in coastal ecology: Studies from Kenya,* pp. 233-242. Leiden: African Studies Centre.

GELCICH, S., G. EDWARDS-JONES & M.J. KAISER (2005). Importance of attitudinal differences among artisanal fishers towards co-management and conservation of marine resources. *Conservation Biology* 19: 865-875.

GLAESEL, H. (1997a). *Fishers, parks, and power: The socio-environmental dimensions of marine resource decline and protection on the Kenya Coast.* (Ph.D. Thesis). Madison: University of Wisconsin.

— (1997b). They are not just 'samaki': Towards an understanding of fisher vocabulary on the Kenya Coast. *Afrikanistische Arbeitspapiere* 51: 165-179.

— (2000). State and local resistance to the expansion of two environmentally harmful marine fishing techniques in Kenya. *Society & Natural Resources* 13: 321-338.

GUARD, M. & M. MASAIGANAH (1997). Dynamite fishing in Southern Tanzania, geographical variation, intensity of use and possible solutions. *Marine Pollution Bulletin* 34 (10): 758-762.

HAUCK, M. & M. SOWMAN (2001). Coastal and fisheries co-management in South Africa: an overview and analysis. *Marine Policy* 25: 173-185.

HEMPHILL, S. (2008). Tuna fisheries: A management conundrum. *Samaki News* (Magazine of the Department of Fisheries Kenya) 5 (1): 17-18.

HERSOUG, B., S. JENTOFT & P. DEGNBOL (2004). *Fisheries development: The institutional challenge.* Delft: Eburon.

HINRICHSEN, D. (1998). *Coastal waters of the world: Trends, threats and strategies.* Washington: Island Press.

HOF, T. VAN HET (1999). *Management plan Malindi and Watamu Marine Parks and Reserves.* Mombasa: Kenya Wildlife Service.

HOORWEG, J. (1998). Introduction. In J. Hoorweg, ed., *Dunes, groundwater, mangroves and birdlife in coastal Kenya.* (Coastal Ecology Series 4). Eldoret: Moi University, School of Environmental Studies.

HOORWEG, J., D. FOEKEN & W. KLAVER (1995). *Seasons and nutrition at the Kenya coast.* (ASC Research Series No.7). Oxford: Avebury.

HOORWEG, J., D. FOEKEN & R. OBUDHO (2000). Conclusion: Culture, resources and development in the Kenya Coast. In J. Hoorweg, D. Foeken & R. Obudho, eds., *Kenya Coast Handbook: Culture, resources and development in the East African littoral,* pp. 393-408. Hamburg: Lit Verlag.

HOORWEG, J., A. DEGEN & B. WANGILA (2003). *Income diversification and resource conservation among artisanal fishers at the central Kenyan coast.* (NIRP End Report). Leiden: African Studies Centre.

HOORWEG, J., N. VERSLEIJEN, B. WANGILA & A. DEGEN (2006). Income diversification and fishing practices among artisanal fishers. Paper presented at Coastal Ecology Conference IV, Mombasa, May 29-30.

HOREMANS, B. & A. JALLOW (1997a). Current state and perspectives of marine fisheries resources co-management in West Africa. In A.K. Norman, J.R. Nielsen & S. Sverdrup-Jensen, eds., *Fisheries co-management in Africa: Proceedings from a regional workshop on fisheries co-management*, pp. 233-254. (Fisheries Co-management Research Project, Research Report No.12). Hirtshals: Institute for Fisheries Management and Coastal Community Development.

— eds. (1997b). *Workshop on gender roles and issues in artisanal fisheries in West Africa, Lome, 11-13 December 1996.* (Programme for the Integrated Development of Artisanal Fisheries in Africa, Technical Report No. 97). Cotonou: Food and Agriculture Organization.

HORRILL, C. & S. MAKOLOWEKA (1998). Silencing the dynamite fisheries along the Tanga Coast, Tanzania. In *International tropical marine ecosystems management symposium proceedings*, pp. 260-266. Townsville, Australia.

IKIARA, G.K. (2000). Employment. In J. Hoorweg, D. Foeken & R. Obudho, eds., *Kenya Coast Handbook: Culture, resources and development in the East African littoral,* pp. 237-246. Hamburg: Lit Verlag.

JACQUET, J.L. & D. ZELLER (2007). Putting the 'United' in the United Republic of Tanzania: Reconstructing marine fisheries catches. In D. Zeller & D. Pauly, eds., *Reconstructing of marine fisheries catches for key countries and regions (1950-2005),* pp. 49-60. Fisheries Research Reports 15 (2). Fisheries Centre, University of British Columbia.

KAAG, M., R. VAN BERKEL, J. BRONS, M. DE BRUIJN, H. VAN DIJK, L. DE HAAN, G. NOOTEBOOM & A. ZOOMERS (2004). Ways forward in livelihood research. In D. Kalb, W. Pansters & H. Siebers (eds), *Globalization and development: Themes and concepts in current research*, pp. 49-74. Dordrecht: Kluwer Academic Publishers.

KAUNDA-ARARA, B. (1997). Analysis of fish catch data from 1985-1994 in the Kenyan inshore marine waters. *African Journal of Tropical Hydrobiology and Fisheries* 7 (1/2): 1-6.

KAUNDA-ARARA, B. & M.J. NTIBA (1997). The reproductive biology of *Lutjanus fulviflamma* (Forsskäl, 1775) (Pisces-Lutjanidae) in Kenyan inshore marine waters. *Hydrobiologia* 353: 153-160.

KAUNDA-ARARE, B., G.A. ROSE, M.S. MUCHIRI & R. KAKA (2003). Long-term trends in coral reef fish yields and exploitation rates of commercial species from coastal Kenya. *Western Indian Ocean Journal of Marine Science* 2: 105-116.

KAUNDA-ARARA, B. & G.A. ROSE. (2004a). Out-migration of tagged fish from marine reef National Parks to fisheries in coastal Kenya. *Environmental Biology of Fishes* 70: 363-372.

— (2004b). Effects of marine reef National Parks on fishery CPUE in coastal Kenya. *Biological Conservation* 118: 1-13.

KENYA (1928). *Report on a preliminary survey of the sea fisheries of Kenya Colony.* (Prepared by C. von Bonde). Nairobi: Colony and Protectorate of Kenya.

— (1938). *Report of the committee on the control and development of fshing in Kenya.* Nairobi: Colony and Protectorate of Kenya.

— (1947). *Review of Kenya Fisheries: 1939 to end of 1945.* Nairobi: Colony and Protectorate of Kenya.

— (1950a). *Review of Kenya Fisheries: 1946-1947.* Nairobi: Colony and Protectorate of Kenya.

— (1950b). *Review of Kenya Fisheries: 1948 and 1949.* Nairobi: Colony and Protectorate of Kenya.

— (1951). *Review of Kenya Fisheries 1950.* Nairobi: Colony and Protectorate of Kenya.

— (1952). *Review of Kenya Fisheries 1951.* Nairobi: Colony and Protectorate of Kenya.

— (1954). *Review of Kenya Fisheries 1953.* Nairobi: Colony and Protectorate of Kenya.

— (1955). *Review of Kenya Fisheries 1954.* Nairobi: Colony and Protectorate of Kenya.

— (1956). *Review of Kenya Fisheries 1955.* Nairobi: Colony and Protectorate of Kenya.

— (1957). *Report on Kenya Fisheries 1956.* Nairobi: Colony and Protectorate of Kenya.

— (1959). *Report on Kenya Fisheries 1958.* Nairobi: Colony and Protectorate of Kenya.

— (1960). *Report on Kenya Fisheries 1959.* Nairobi: Colony and Protectorate of Kenya.

— (1962). *Report on Kenya Fisheries 1961.* Nairobi: Colony and Protectorate of Kenya.

— (1963). *Report on Kenya Fisheries 1962.* Nairobi: Government of Kenya.

— (1964a). *Report on Kenya Fisheries 1963.* Nairobi: Government of Kenya.

— (1964b). *Development plan 1964-1970.* Nairobi: Government of Kenya.

— (1966). *Report on Kenya Fisheries 1964.* Nairobi: Republic of Kenya.

— (1967). *Report on Kenya Fisheries 1965.* Nairobi: Republic of Kenya.

— (1968). *Report on Kenya Fisheries 1966.* Nairobi: Republic of Kenya.

— (1969). *Development plan 1970-1974.* Nairobi: Republic of Kenya.
— (1974). *Development plan 1974-1978; Vol. I-II.* Nairobi: Republic of Kenya.
— (1975). *Statistical abstract 1975.* Nairobi: Central Bureau of Statistics.
— (1989). *The Wildlife (Conservation and Management) Act. Chapter 376.* Nairobi: Government Printer.
— (1991). *Fisheries act. Chapter 373.* Laws of Kenya. Nairobi: Government Printer.
— (1997). *Malindi district development plan.* Nairobi: Government Printer.
— (1999a). *The Maritime Zones Act. Chapter 371.* (www.lexadin.nl accessed on 24.08.2008).
— (1999b). *The Environmental Management and Coordination Act* (www. lexadin.nl accessed on 24.08.2008).
— (2001). *Poverty reduction strategy paper for the period 2001-2004.* Nairobi: Republic of Kenya, Ministry of Finance and Planning.
— (2002a). *Kenya 1999 population and household census: The popular report.* Nairobi: Central Bureau of Statistics.
— (2002b). *Kenya 1999 population and household census: Vol. VII, Analytical Report on Population Projections.* Nairobi: Central Bureau of Statistics.
— (2003). *Statistical abstract 2003.* Nairobi: Central Bureau of Statistics.
— (2005). *Geographic dimensions of well-being in Kenya: Who and where are the poor? Vol. II.* Nairobi: Central Bureau of Statistics
— (2007a). *Statistical abstract 2007.* Nairobi: Central Bureau of Statistics.
— (2007b). *Draft Sessional Paper on National Fisheries Policy.* Nairobi: Republic of Kenya.
KING, A. (2003). Strategies used by local fishers to ensure access to and control over scarce resources in Galu and the wider implications for marine resource management. In J. Hoorweg & N. Muthiga, eds., *Recent advances in coastal ecology: Studies from Kenya,* pp. 215-232. Leiden: African Studies Centre.
KLAVER, W. & R. MWADIME (2000). Food consumption and nutrition. In J. Hoorweg, D. Foeken & R. Obudho, eds., *Kenya Coast Handbook: Culture, resources and development in the East African littoral,* pp. 279-295. Hamburg: Lit Verlag.
KLEIN, A. (1999). The Barracuda's tale: Trawlers, the informal sector and a state of classificatory disorder off the Nigerian Coast. *Africa* 69 (4): 555-573.
KNOWLES, T. (1987). Kenya's muslim maritime community. *Cultural survival quarterly,* 11, 2.
KOCHEY, J. (2008). Prawn trawling in Ungwana Bay, Malindi, Kenya. *Samaki News* (Magazine of the Department of Fisheries Kenya) 5 (1): 19.
KOFFSKI, P.L. (1977). *History of Takaungu East Africa.* (Ph.D. thesis). Madison: University of Wisconsin.
LOPES, S., E. POIOSSE, J. WILSON, J.L. KROMER, L. MANUEL, C. CULULO & M.A.R. PINTO (1997). From no management towards co-management: A case study on artisanal fisheries in Angoche district, Nampula Province, Mozambique. In A.K. Norman, J.R. Nielsen & S. Sverdrup-Jensen, eds., *Fisheries co-management in Africa: Proceedings from a regional workshop on fisheries co-management.* (Fisheries Co-management Research Project, Research Report No.12). Hirtshals: Institute for Fisheries Management and Coastal Community Development.
MAGERIA, C., O. MAKOGOLA & S. NDEGWA (2008). Current status of Kenyan coastal fisheries. *Samaki News* (Magazine of the Department of Fisheries Kenya) 5 (1): 6-7.

MAKOLOWEKA, S. & K. SHURCLIFF, K. (1997). Coastal management in Tanga, Tanzania: A decentralized community-based approach. *Ocean and Coastal Management* 37 (3): 349-357.

MALLERET-KING D (2003). Food security benefits of the Kisite Marine National Park for the surrounding fishing communities. In J. Hoorweg & N. Muthiga, eds., *Recent advances in coastal ecology: Studies from Kenya,* pp. 201-214. Leiden: African Studies Centre.

MALLERET-KING, D., A. KING, S. MANGUBHAI, J. TUNJE, J. MUTURI, E. MUENI & H. ON'GANDA (2003). *Understanding fisheries associated livelihoods and the constraints to their development in Kenya and Tanzania: Livelihood appraisal.* (FMSP Project R8196). DFID: Fisheries Management Science Programme.

MANGI, S.C. & C.M. ROBERTS (2006). Quantifying the environmental impacts of artisanal fishing gear on Kenya's coral reef ecosystems. *Marine Pollution Bulletin* 52: 1646-1660.

— (2007). Factors influencing fish catch levels on Kenya's coral reefs. *Fisheries Management & Ecology* 14 (4): 245-253.

MARTIN, E.B. (1973). *The history of Malindi: A geographical analysis of an East African coastal town from the Portuguese period to the present.* Nairobi: East African Literature Bureau.

MASAI, W.S. (2000). Industrialization. In J. Hoorweg, D. Foeken & R. Obudho, eds., *Kenya Coast Handbook: Culture, resources and development in the East African littoral,* pp. 211-221. Hamburg: Lit Verlag.

MASALU, D.C.P. (2000). Coastal and marine resource use conflicts and sustainable development in Tanzania. *Ocean & Coastal Management* 43: 475-494.

MATHOOKO, J.M. (2005). Application of traditional ecological knowledge in the management and sustainability of fisheries in East Africa. *Hydrobiologia* 537: 1-6.

MCCLANAHAN, T.R. (1988). Seasonality in East Africa's coastal waters. *Marine Ecology, Progress Series* 44: 191-199.

— (1996). Oceanic ecosystems and pelagic fisheries. In T.R. McClanahan & T.P. Young, eds., *East African ecosystems and their conservation*, pp. 39-66. Oxford: Oxford University Press.

MCCLANAHAN, T.R. & D. OBURA (1995). Status of Kenyan coral reefs. *Coastal Management* 23: 57-76.

— (1996). Coral reefs and nearshore fisheries. In T.R. McClanahan & T.P. Young, eds., *East African ecosystems and their conservation*, pp. 67-99. Oxford: Oxford University Press.

MCCLANAHAN, T.R., H. GLAESEL, J. RUBENS & R.W. KIAMBO (1997). The effects of traditional fisheries management on fisheries yields and the coral-reef ecosystems of southern Kenya. *Environmental Conservation* 24 (2): 105-120.

MCCLANAHAN, T.R. & S. MANGI (2000). Spillover of exploitable fishes from a Marine Park and its effect on the adjacent fishery. *Ecological Applications* 10 (6): 1792-1805.

MCCLANAHAN, T.R., J. DAVIES & J. MAINA (2005a). Factors influencing resource users and managers' perceptions towards marine protected area management in Kenya. *Environmental Conservation* 32 (1): 42-49.

MCCLANAHAN, T.R., J. MAINA & J. DAVIES (2005b). Perceptions of resource users and managers towards fisheries management options in Kenyan coral reefs. *Fisheries Management & Ecology* 12: 105-112.

McClanahan, T.R., S. Mwaguni & N.A. Muthiga (2005c). Management of the Kenyan coast. *Ocean & Coastal Management* 48: 901-931.

MENR (1994). *The Kenya national environment action plan*. Nairobi: Ministry of Environment and Natural Resources.

Middleton, J. (2000). The peoples. In J. Hoorweg, D. Foeken & R. Obudho, eds., *Kenya Coast Handbook: Culture, resources and development in the East African littoral,* pp. 101-114. Hamburg: Lit Verlag.

Mohammed, M.O. (2002). *Fish catch composition and some aspects of reproductive biology of Siganus sutor along the Malindi-Kilifi marine inshore waters.* (M.Phil thesis). Eldoret: Moi University, Department of Fisheries.

MRAG (2005). *Review of impacts of illegal, unreported and unregulated fishing on developing countries (Final Report).* London: Marine Resources Assessment Group.

Mwadime, R. (1996). *Non-farm employment in rural Kenya: Micro-mechanisms influencing food and nutrition of farming households.* (Ph.D. thesis). Wageningen: Wageningen Agricultural University.

Mwakilenge, E. (1996). *Towards sustainable development in the fishing industry of Coast Province.* (Technical Report 1). Eldoret: Moi University, Dept. of Fisheries.

Neiland, A.E. (2004). Fisheries development, poverty alleviation and small-scale fisheries: A review of policy and performance in developing countries since 1950. In A.E. Neiland & C. Béné, eds., *Poverty and small-scale fisheries in West Africa*, pp. 189-208. Dordrecht: Kluwer Academic Publishers.

Niemeijer, R., D. Foeken & W. Klaver (1991). *Seasonality in the coastal lowlands of Kenya, Part 4/5: Food consumption and anthropometry.* Nairobi/Leiden: Ministry of Planning and National Development/African Studies Centre, Food and Nutrition Studies Programme. (Report 38).

Ntiba, M.J. (1986). *The biology of the Kenyan reef fish of the genus Siganus.* (M.Sc. thesis.) University of Nairobi.

Nzioka, R.M. (1981). *Biology and fishery of the Scolopsis bimaculatus (Rüppel) in Kenya.* (M.Sc. thesis). University of Nairobi.

— (1984). The evaluation of marine fisheries resources in Kenya. In S.A. Iversen & S. Myklerou, eds, *NORAD-Kenya seminar to review marine fish and stocks in Kenya,* pp. 27-32. Mombasa: Kenya Marine and Fisheries Research Institute.

Obura, D. (2000). Status report Kenya. In O. Lindén & N. Sporrong, eds., *Coral reef degradation in the Indian Ocean: Status report and project presentations 1999,* pp. 33-36. Mombasa: CORDIO/Stockholm: Stockholm University, SAREC Marine Science Program, Department of Zoology,

— (2001). Kenya. *Marine Pollution Bulletin* 42 (12): 1264-1278.

Obura, D.O., S. Wells, J. Church & C. Horrill (2002). Monitoring of fish and fish catches by local fishermen in Kenya and Tanzania. *Marine and Freshwater Research* 53: 215-222.

Ochieng, B, J. Cappon, J. Mbendo & B. Makau (2001). *Kenya coastal management initiative: Assessment of laws, policies and institutions.* Nairobi: African Centre for Technology Studies.

Ochiewo, J. (2004). Changing fisheries practices and their socio-economic implications in South Coast Kenya. *Ocean & Coastal Management* 47: 389-408.

Odero, N. (1984). Marine fisheries development in Kenya. In S.A. Iversen & S. Myklerou, eds, *NORAD-Kenya seminar to review marine fish and stocks in Kenya,* pp. 11-14. Mombasa: Kenya Marine and Fisheries Research Institute.

OKEMWA, E.N., R.K. RUWA & B.A.J. MWANDOTTO (1997). Integrated coastal zone management in Kenya: Initial experiences and progress. *Ocean and Coastal Management* 37 (3): 319-347.

OKIDI, C.O. (1979). *Kenya's marine fisheries: An outline of policy and activities.* (Occasional Paper 30). Nairobi: Institute for Development Studies.

OKIDI, C.O. & S. WESTLEY (1978). *Management of coastal and offshore resources in eastern Africa.* (IDS Occasional Paper 28). Nairobi: Institute for Development Studies.

OMMANNEY, F.D. (1955). *Isle of cloves: A view of Zanzibar.* London: Longmans & Green.

OSTROM, E. (1999). Coping with tragedies of the commons. *Annual Review of Political Science* 2: 493-535.

PAICE, E. (2001). *Lost lion of empire: The life of 'Cape-to-Cairo' Grogan.* New York: Harper/Collins.

PAULY, D., V. CHRISTENSE, S. GUÉNETTE, T.J. PITCHER, U.R. SUMAILA, C.J. WALTERS, R. WATSON & D. ZELLER (2002). Towards sustainability in world fisheries. *Nature* 418: 689-695.

PAYNE, I. (2000). The changing role of fisheries in development policy. *Natural Resource Perspectives* 59 (June 2000). London: Overseas Development Institute.

PLUMMER, R. & D. FENNEL (2007). Exploring co-management theory: Prospects for sociobiology and reciprocal altruism. *Journal of Environmental Management* 85: 944-955.

POMEROY, R.S. & F. BERKES (1997). Two to tango: The role of government in fisheries co-management. *Marine Policy* 21 (5): 465-480.

POMEROY, R.S. & R. RIVERA-GUIEB (2006). *Fishery co-management: A practical handbook.* Wallingford: CAB International.

RICHMOND, M.D., ed. (1997). *A guide to the seashores of Eastern Africa and the Western Indian Ocean Islands.* Zanzibar: SIDA.

ROBERTS, C., J. BOHNSACK, F. GELL, J. HAWKINS & R. GOODRIDGE (2001). Effects of marine reserves on adjacent fisheries. *Science* 294: 1920-1923.

SCOONES, I. (1998). *Sustainable rural livelihoods: A framework for analysis.* (IDS Working Paper 72). Brighton: Institute of Development Studies.

SMITH, M.M. & P.C. HEEMSTRA, eds. (1991). *Smith's sea fishes.* Johannesburg: Southern Book Publishers.

STATOIDS (2008). Malindi district & Kilifi district. (www.statoids.com)

TOURAY, I. (1996). Gender issues in the fisheries sector and effective participation. In B.W. Horemans & A.M. Jallow, eds., *Report of the workshop on gender roles and issues in artisanal fisheries in West Africa, Lome, December 1996.* Cotonou: DANIDA & FAO.

TRAVIS, W. (1959). *Beyond the reefs.* London: George Allen & Unwin.

— (1961). *Sharks for sale.* London: George Allen & Unwin.

— (1967). *The voice of the turtle.* London: George Allen & Unwin.

TUNJE, J.G. (2000). *Reef fisheries in Kilifi and Lamu districts: Fishing practices, awareness of resource degradation and traditional ways of conservation among artisanal fishermen.* (M.Phil. thesis). Eldoret: Moi University, School of Environmental Studies.

TUNJE, J.G. & J. HOORWEG (2001). Fishing practices in protected and non-protected marine areas of Kenya. Paper presented at the Conference on People and the Sea: Maritime research in the social sciences. Amsterdam, Aug-Sep. University of Amsterdam: Centre for Maritime Research.

— (2003). Awareness of resource degradation among artisanal fishermen in Kilifi and Lamu. In J. Hoorweg & N. Muthiga, eds., *Recent advances in coastal ecology: Studies from Kenya*, pp. 185-200. Leiden: African Studies Centre

TVEDTEN, I. & B. HERSOUG (1992). Introduction. In I. Tvedten & B. Hersoug, eds., *Fishing for development: Small-scale fisheries in Africa,* pp. 7-30. Uppsala: Nordiska Afrikainstitutet.

UNDP (2000). *Human development report 2000.* New York: Oxford University Press/United Nations Development Programme.

UNEP (1998a). *Eastern Africa atlas of coastal resources 1: Kenya.* Nairobi: United Nations Environment Programme.

— (1998b). *Regional seas: Management and conservation of renewable resources in the South Asian seas region.* (UNEP Regional Seas Report and Studies. No. 62). Nairobi: United Nations Environment Programme.

VENEMA, S.C. (1984). Review of marine resource surveys in Kenyan waters. In S. Iversen & S. Myklevoll, eds., *The proceedings of the Norad-Kenya seminar to review the marine fish stocks and fisheries in Kenya, Mombasa, March 13-15*, pp. 61-82. Mombasa/Bergen: Kenya Marine and Fisheries Research Institute/Institute of Marine Research.

VERSLEIJEN, N. (2001). *An empty sufuria: The effects of a Marine National Park on the livelihood strategies and income diversification of fishermen household at the Kenya Coast.* (MA Thesis). Wageningen University and Research Centre.

VERSLEIJEN, N. & J. HOORWEG (2006). Marine conservation: The voice of the fishers. Paper presented at Coastal Ecology Conference IV, Mombasa, May 29-30.

— (2008). From farming to fishing: Marine resource conservation and a new generation of fishermen. *Western Indian Ocean Journal of Marine Science* 7 (1): 1-14.

WAKAJUMMAH, J.O. (2000). Population dynamics. In J. Hoorweg, D. Foeken & R. Obudho, eds., *Kenya Coast Handbook: Culture, resources and development in the East African littoral,* pp. 73-83. Hamburg: Lit Verlag.

WALMSLEY, S., J. PURVIS & C. NINNES (2006). The role of small-scale fisheries management in the poverty reduction strategies in the Western Indian Ocean region. *Ocean and Coastal Management* 49: 812-833.

WAMUKOTA, A.W. (2002). *Processing and marketing of fish among the coastal fisher-folk.* (M.Phil thesis). Moi University: Department of Environmental Economics.

WAMUKOYA, G.M., J.M.P. MIRANGI & W.K. OTTICHILO (1996). *Report on the marine aerial survey of marine mammals, sea turtles, sharks and rays.* (KWS Technical Series Report No. 1). Mombasa: Kenya Wildlife Service.

WANGILA, B.C.C., J.C. HOORWEG & A. DEGEN (2007a). Fishing in the neigbourhood of a Marine Protected Area: The case of Central Coast of Kenya artisanal fishery. *Journal of Science Technology Education and Management* 1 (2): 46-60.

— (2007b). Gender analysis of fish traders in the Central Coast inshore artisanal fishery of Kenya. *Journal of Science Technology Education and Management* 1 (1): 42-54.

WHAT (2000). *Governance for sustainable future. II: Fishing for the future,* London: World Humanity Action Trust.

WILLIAMS, M.J., N.H. CHAO, P.S. CHOO, K. MATICS, M.C. NANDEESHA, M. SHARIFF, I. SIASON, E. TECH E & J.M.C. WONG, eds. (2002). *Global symposium on women in fisheries: Sixth Asian fisheries forum.* Penang: World Fish Center (www.worldfishcenter.org > scientific publications > 2003-2000)

WILLMAN, R. (2004). Poverty in coastal fishing communities. In A.E. Neiland & C. Béné, eds., *Poverty and small-scale fisheries in West Africa*, pp. 245-252. Dordrecht: Kluwer Academic Publishers.

WILSON, D.C., J.R. NIELSEN & P. DEGNBOL, eds. (2003). *The fisheries co-management experience: Accomplishments, challenges and prospects.* Dordrecht: Kluwer Academic Publishers.

WIOMSA (2007). Marine protected areas of the Western Indian Ocean. (http://www.wiomsa.org/mpatoolkit/Themesheets/MPA_details.pdf (accessed 29.09.2007)

WOODHOUSE, P. (2002). *Natural resource management and chronic poverty in Sub-Saharan Africa: An overview paper.* (CPRC Working Paper 14). Manchester: University of Manchester, Institute for Development, Policy and Management.

WRIGHT, S. (2008). Kenyan seasonal catch results. *Rainbow Runner* (Magazine of the Kenya Association of Sea Anglers), 36-49

Index

www.ingramcontent.com/pod-product-compliance
Lightning Source LLC
Chambersburg PA
CBHW071345230926
51846CB00002B/449
9789004178083